I0808143

Seaforth

WORLD NAVAL REVIEW

2022

LHD 6

Seaforth

WORLD NAVAL REVIEW 2022

Editor
CONRAD WATERS

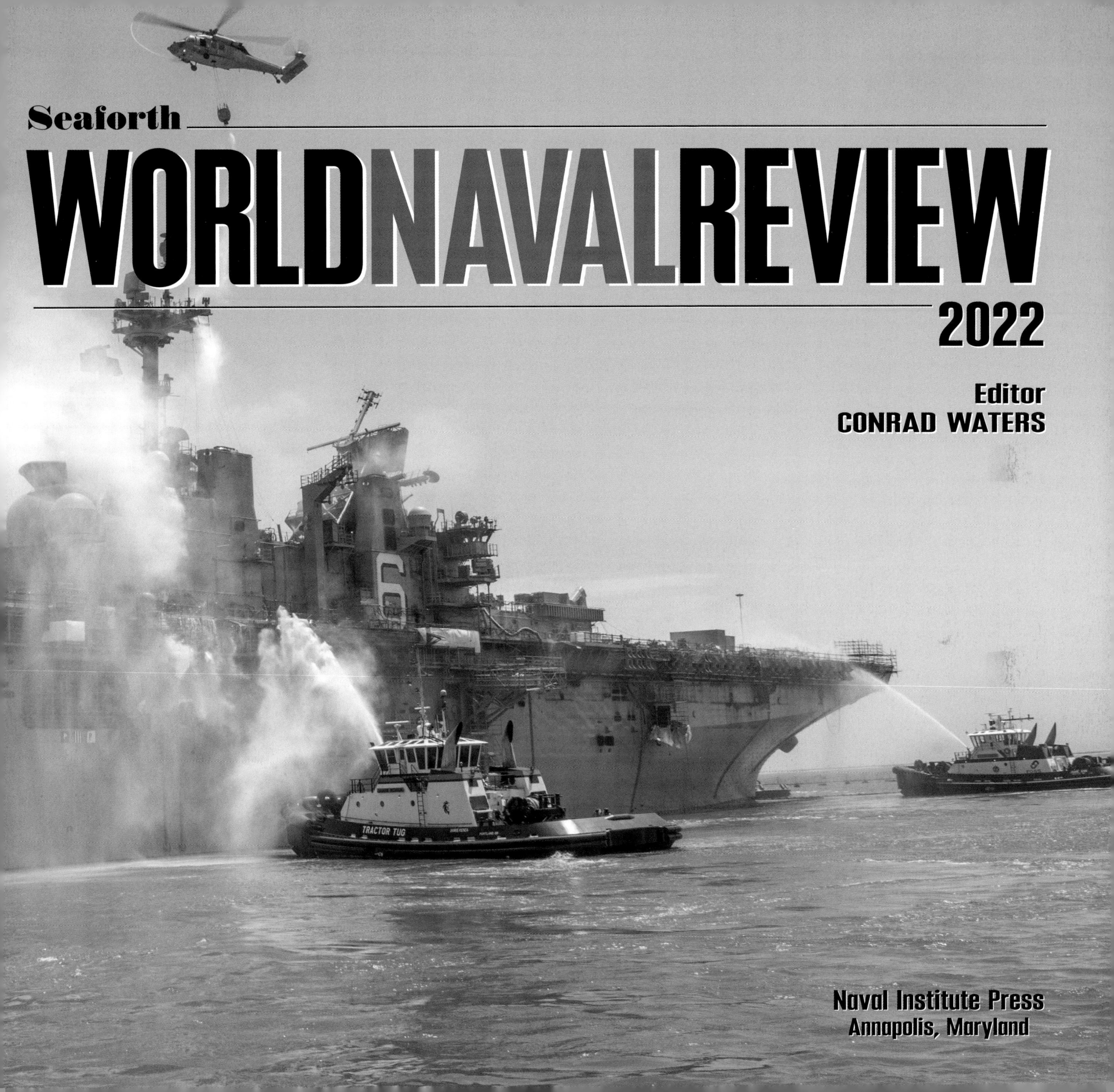

Naval Institute Press
Annapolis, Maryland

Frontispiece: Firefighters struggle to contain a major blaze aboard the US Navy's amphibious assault ship *Bonhomme Richard* (LHD-6) that broke out whilst the ship was alongside at San Diego in the morning of Sunday 12 July 2020. The ship – newly converted to support F-35B operations – was ultimately declared a total loss and disposed of by scrapping. Her destruction represents a major blow to a fleet that is already struggling to maintain adequate ship numbers. *(US Navy)*

First published in Great Britain in 2021 by
Seaforth Publishing
An imprint of Pen & Sword Books Ltd
47 Church Street, Barnsley
S Yorkshire S70 2AS

www.seaforthpublishing.com
Email info@seaforthpublishing.com

Published and distributed in the United States of America and Canada by the Naval Institute Press,
291 Wood Road, Annapolis, Maryland 21402-5043
www.nip.org

Library of Congress Control Number: 2021946140

ISBN 978 1 68247 746 5

Typeset and designed by Stephen Dent
Printed and bound in India by Replika Press Pvt. Ltd

FSC www.fsc.org MIX Paper from responsible sources FSC® C016779

CONTENTS

Section 1: Overview

Section 2: World Fleet Reviews

Section 3: Significant Ships

Section 4: Technological Reviews

Note on Tables: Tables are provided to give a broad indication of fleet sizes and other key information but should be regarded only as a general guide. For example, many published sources differ significantly on the principal particulars of ships, whilst even governmental information can be subject to contradiction. In general terms, the data contained in these tables is based on official information updated as of June 2021, supplemented by reference to a wide range of secondary and corporate sources, such as shipbuilder websites.

1 OVERVIEW
INTRODUCTION

Author:
Conrad Waters

'Divide and rule, a sound motto. Unite and lead, a better one', wrote the German polymath Johann Wolfgang von Goethe (1749–1832), considered by many to be that nation's greatest writer of the period. His words have resonance in the current day as the United States' new Biden administration attempts to make a clean break with the 'America First' nationalism of the Trump era. In contrast, its own security strategy places renewed emphasis on collaboration with like-minded democracies to secure the country's interests. This new approach was evidenced in the United States' new *Interim National Security Strategic Guidance* released on 3 March 2021.[1] Opening with the statement, '… one thing is certain: we will only succeed in advancing American interests and upholding our universal values by working in common cause with our closest allies and partners …' the document emphasises collective action with friendly nations to protect the rules-based international order that is seemingly under siege from authoritarian powers.

Whilst the mood music at the top may have changed, it will likely take considerable time and effort for the United States to regain the trust of traditional allies. Confidence in American reliability has been shaken by the transactional approach to relationships that typified the Trump years, perhaps best illustrated by the often publicly voiced criticism of the 'obsolete' NATO alliance on grounds of inadequate burden-sharing. One early bump along the road to recovery was President Biden's decision to set 11 September 2021 as a firm date to withdraw all American forces from Afghanistan. This step was seemingly taken without full consultation with the NATO and other allied nations who have assisted military operations in the country over the past two decades and has drawn criticism from many quarters.

The Portuguese MEKO 200 frigate *Corte-Real* pictured operating with Danish and Swedish warships in the course of the Trident Juncture 2018 exercise. The new American Biden administration is attempting to rebuild alliances with NATO and other friendly nations to counter growing rivalry with Russia and China. *(Marius Vågenes Villanger / Norwegian Armed Forces)*

In truth, the withdrawal from Afghanistan is yet further evidence of a reshaping of US strategic priorities to counter the threats arising from China – 'the only competitor capable of combining its power … to mount a sustained challenge to a stable and open international system' – and a disruptive Russia. Lip-service continues to be paid to the other dangers to American security previously identified; viz. the challenges posed by North Korea, Iran and international terrorism. However, the trend towards placing China and Russia at the heart of American security concerns is confirmed. A Global Posture Review mandated by the current administration will emphasise this reality, bolstering the United States' presence in the Indo-Pacific and Europe at the expense of other regions. This shift is likely to increase the importance of 'high end' naval capabilities relative to the largely land-based counter-insurgency forces that fought the long war against terror.

Two main questions need to be answered in assessing whether the United States' strategic shift is likely to be successful. One is the extent to which America's allies are willing to join with it in challenging China's emerging power. An example of this uncertainty was provided in the long-awaited conclusions to the United Kingdom's Integrated

Review of foreign policy, defence, security and international development, finally published on 16 March 2021.[2] Whilst this singled out Russia as 'the most acute threat' to British interests, the stance taken towards China's emerging power was more nuanced. This essentially reflected the inherent conflict that exists between China's important role as a global trading partner and investor and its less beneficial desire to use its economic might to challenge democratic 'Western' values. The United Kingdom is seemingly willing to include a flag-waving transit of the South China Sea on the *Queen Elizabeth* carrier strike group's maiden deployment, also tweaking Russia's tail off the Crimea in the course of the deployment. However, it remains to be seen how firm and widespread the resolve to prioritise intangible principles over hard economic interests will prove to be.

The other question is whether the 'Western' nations will be willing to shoulder the expense of the increased military burden that responding to a new age of great-power competition will inevitably demand. The Biden presidency's own stance in this regard is less than straightforward, as a desire to fund social and economic programmes means that defence is being accorded a lower priority than under the previous administration. The President's FY2022 Defense Budget Request announced on 28 May 2021 – which is essentially flat in real terms – was notable in cutting back on previously planned naval shipbuilding programmes that are a key aspect of countering growing Chinese naval power. Nevertheless, many other allied countries are increasing defence spending, the financial damage of the COVID-19 pandemic notwithstanding. Alongside investment in new capabilities such as cyber, unmanned and even space capabilities, much of the additional spending is being directed towards making good gaps in naval capacity.

DEFENCE BUDGETS AND PLANS

The shift towards expanding defence budgets in spite of the impact of the current pandemic is demonstrated by the latest annual assessment provided by the independent Stockholm International Peace Research Institute (SIPRI). Published in April 2021 and updating spending data up to 2020, this saw expenditure rise from 2.2 percent to 2.4 percent of GDP or global economic output year-on-year. This is broadly equivalent to the percentage figure seen ten years previously and confirmed a recent reversal in a trend that had previously seen the so-called 'world military burden' steadily shrink for much of the decade. The main underlying driver of this higher military burden was the reduction in output resulting from severe global economic downturn caused by the pandemic. However, defence spending did actually rise by some 2.6 percent in real terms. Whilst commentators are divided as to whether stretched government finances will allow this uplift in spending to be sustained over the medium term, plans that have already been announced for 2021–2022 suggest defence expenditure will continue to grow in the immediate future.

Table 1.1: COUNTRIES WITH HIGH NATIONAL DEFENCE EXPENDITURES: 2020

RANK 2020	(2019)	COUNTRY	TOTAL US$BN 2020[1]	SHARE OF GDP 2020	WORLD SHARE 2020	TOTAL US$BN 2019[1]	REAL CHANGE 2019–2020[2]	REAL CHANGE 2011–2020[2]
1	(1)	United States	778	3.7%	39%	734	4.4%	-10%
2	(2)	China	[252]	[1.7%]	[13%]	[240]	[1.9%]	[76%]
3	(3)	India	72.9	2.9%	3.7%	71.5	2.1%	34%
4	(4)	Russia	61.7	4.3%	3.1%	65.2	2.5%	26%
5	(6)	United Kingdom	59.2	2.2%	3.0%	56.9	2.9%	-4.2%
6	(5)	Saudi Arabia	[57.5]	[8.4%]	[2.9%]	[62.0]	[-10.0%]	[2.3%]
7	(8)	Germany	52.8	1.4%	2.7%	49.0	5.2%	28%
8	(7)	France	52.7	2.1%	2.7%	50.1	2.9%	9.8%
9	(9)	Japan	49.1	1.0%	2.5%	47.6	1.2%	2.4%
10	(10)	South Korea	45.7	2.8%	2.3%	43.9	4.9%	41%
-	(-)	**World Total**	**1981**	**2.4%**	**100%**	**1909**	**2.6%**	**9.3%**

Information from the Stockholm International Peace Research Institute (SIPRI) – https://www.sipri.org/databases/milex
The SIPRI Military Expenditure Database contains data on countries over the period 1949-2020.

Notes:

1. US$ totals for 2020 and 2019 are based on then current (i.e. non-inflation adjusted) prices and exchange rates for the years in question. Exchange rate movements, in particular, can therefore result in significant movements in the US$ figures and explain apparent discrepancies in the table. SIPRI also adjust previous-year calculations when more accurate data becomes available – this has resulted in several significant differences from the 2019 total figures contained in last year's *Seaforth World Naval Review*. Notably, spending for the United Kingdom was revised c. 10 percent upwards due to the inclusion of certain spending on pensions and other items, as well as taking into account income generated by the MoD that was then recirculated into military spending.

2. The 'real' change figure is based on constant (2019-based) US$ figures.

3. Figures in brackets are SIPRI estimates.

Table 1.1 lists the main defence spending nations as of 2020. All in all, the data showed little in the way of material change in the distribution of budgets compared with 2019. Notably, there was no alteration to the constituents of the list of the top ten spending countries which – in total – accounted for an unchanged three-quarters of overall global military expenditure. Perhaps unsurprisingly, eight of these ten can also be considered as major naval powers. Other countries with significant fleets – namely Italy, Australia, Canada and Brazil – all ranked between 11 and 15 in the major spending league.

The United States continued to have by far the largest military budget in the world, albeit a final 4.4 percent uplift from the Trump years still left spending down by around 10 percent from a decade ago. The United Kingdom – now ranked in fifth place following a material upward revision to how SIPRI calculates British defence spending – remained the only other country to see a real-terms decline in spending over this period, albeit the trend is now being reversed.[3] As previously, it was countries within the broader Asian region that demonstrated the most sustained rise in military expendi-

The People's Liberation Army Navy Type 054A frigate *Anyang* leads the Type 052C destroyer *Jinan* and the newer Type 052D variant *Kunming* in exercises in the East China Sea on 23 April 2021. China is building warships at a pace that rival navies are struggling to match. *(China Military Online)*

ture. China, South Korea and India – as well as Australia outside the top ten – have all seen spending grow by a third or more since 2011. Moreover, China and India are both firmly ensconced as the second and third highest spenders, a status reinforced by the fact that their budgets tend to buy far more in local currency than those of their counterparts in the 'West'.

FLEET REVIEWS

An assessment of Table 1.2 summarising the year-on-year development of major fleet strengths provides an insight into some of the practical realities that might be influencing the Biden administration's focus on rebuilding the United States' alliances. The data shows that the US Navy remains the world's most powerful and balanced fleet. However, its main rival – China's People's Liberation Army Navy (PLAN) – is continuously closing the gap. Indeed, despite the relatively generous funding the US Navy benefitted from during the Trump

TABLE 1.2: MAJOR FLEET STRENGTHS 2020-2021[1]

REGION	THE AMERICAS				EUROPE & RUSSIA										ASIA								IND. OCEAN	
COUNTRY	USA		BRAZIL		UK		FRANCE		ITALY		SPAIN		RUSSIA		CHINA		JAPAN		KOREA(S)		AUSTRALIA		INDIA	
	2020	2021	2020	2021	2020	2021	2020	2021	2020	2021	2020	2021	2020	2021	2020	2021	2020	2021	2020	2021	2020	2021	2020	2021
Carriers & Amphibious																								
CV/CVN	11	11	–	–	2	2	1	1	1	1	–	–	1	1	2	2	–	–	–	–	–	–	1	1
CVS/CVH	–	–	–	–	–	–	–	–	1	1	–	–	–	–	–	–	4	4	–	–	–	–	–	–
LHA/LHD/LPH	10	9	1	1	–	–	3	3	–	–	1	1	–	–	–	1	–	–	1	2	2	2	–	–
LPD/LSD	23	22	1	1	5	5	–	–	3	3	2	2	–	–	6	8	3	3	–	–	1	1	1	1
Submarines																								
SSBN	14	14	–	–	4	4	4	4	–	–	–	–	12	12	7	7	–	–	–	–	–	–	1	1
SSN/SSGN	56	54	–	–	7	6	5	6	–	–	–	–	25	25	9	9	–	–	–	–	–	–	1	–
SSK	–	–	5	5	–	–	–	–	8	8	2	2	20	20	50	50	20	21	18	18	6	6	14	14
Surface Combatants																								
BB/BC	–	–	–	–	–	–	–	–	–	–	–	–	2	2	–	–	–	–	–	–	–	–	–	–
CG/DDG/FFG	92	93	7	7	19	18	15	15	16	15	11	11	30	35	70	80	38	36	23	24	11	11	23	22
DD/FGS/FS	22	24	2	2	–	–	13	12	–	–	–	–	40	45	60	75	6	6	12	10	–	–	12	12
FAC[2]	–	–	–	–	–	–	–	–	–	–	–	–	35	30	75	75	6	6	18	18	–	–	8	8
Other (Selected)																								
MCMV	11	8	4	3	13	13	13	13	10	10	6	6	40	40	25	25	21	20	9	9	4	4	–	–
AO/AOR/AFS	29	29	1	1	9	7	3	2	3	4	2	2	20	20	12	12	5	5	3	3	1	2	4	4

Notes

1 Numbers are based on official sources, where available, supplemented by news reports, published intelligence data and other 'open sources' as appropriate. Given significant variations in available data, numbers should be regarded as indicative, particularly with respect to Russia, China and minor warship categories (which are sometimes rounded). There is also a degree of subjectivity with respect to warship classifications given varying national classifications and this can also lead to inconsistency.

2 FAC numbers relate to ships fitted with or for surface-to-surface missiles.

administration, fleet numbers actually fell in a number of important warship categories during the last year.[4] By contrast, major additions to the PLAN over the same period included – amongst other vessels – three major amphibious warships, seven destroyers and around fifteen littoral corvettes. Simply put, the United States is struggling to match the phenomenal rate of naval shipbuilding being achieved by its Asian counterpart that shows little signs of slowing.

Clearly, numbers are not the whole story. The United States Navy continues to enjoy significant – albeit diminishing – qualitative advantages in a number of areas. Combined with new operational concepts such as distributed maritime operations and manned-unmanned teaming, this may be enough to maintain an overall advantage in the years ahead.[5] However, given that America's relative economic and, hence, military and naval power, is likely to continue to decline, the need to place increasing reliance on allied navies to maintain a margin of supremacy is evident. This strategy is not only subject to the political uncertainties already referenced above; it will also need a level of collaboration between the US Navy and friendly fleets not seen since the height of the Cold War to work effectively. On a practical level, therefore, it seems likely that efforts to improve inter-operability and, indeed, integration with friendly navies will be given renewed emphasis in the years ahead.

Two fleets which have long had close links with the US Navy – the British Royal Navy and Spain's *Armada Española* – are subject to detailed review in this edition of *Seaforth World Naval Review.* Richard Beedall's biennial assessment of the British 'Senior Service' paints a broadly positive picture of revived capabilities and ambitious plans, albeit against a backdrop of continued uncertainty over the United Kingdom's longer-term direction. Meanwhile, first-time contributor Alejandro Vilches shows how a combination of American assistance and a long-term focus on developing an indigenous naval shipbuilding industry has created a technologically advanced *Armada* that – in spite of recent financial constraints – has served to place Spain's naval

The Spanish Navy Aegis-equipped frigate *Cristobal Colon* seen in the course of operations with the new US Navy destroyer *Paul Ignatius* (DDG-117) in May 2021. The rebuilding of the Spanish Navy into a technologically advanced force during and after the later stages of the Cold War has owed much to American assistance. *(Crown Copyright 2020)*

The Sri Lanka Navy's 'Colombo' type fast attack craft *P415*. Having played a leading role in a successful counter-insurgency war, the Sri Lanka Navy is now transitioning from 'brown water' warfighting assets towards ships better suited for constabulary missions on the high seas. *(Tetsuya Kakitani)*

The US Navy is replacing its veteran Landing Craft Air Cushion hovercraft with the new ship to shore connector. This picture shows *LCAC-39* during a deployment in Saudi Arabia in March 2021. *(US Navy)*

construction sector firmly on the world stage. A very different type of navy is described by Mrityunjoy Mazumdar. His latest review of the various Indo-Pacific fleets focuses on the Sri Lanka Navy's ongoing transition from a brown water counter-insurgency structure to becoming an ocean-going constabulary force. The fact that the United States, China and India are all supporting these ambitions with the supply of offshore patrol vessels serves as something of a microcosm of these nations' broader maritime rivalries.

SIGNIFICANT SHIPS

The Sri Lanka Navy's reorientation towards constabulary operations is an indication that – away from great power rivalries – it is the mundane but vital duties of ensuring the security and safety of national waters from criminal activity and disaster that is the main *raison d'être* for many navies and coast guards. Indeed, the fire and subsequent loss of the container ship MV *X-Press Pearl* and her cargo of hazardous chemicals off the coast of Colombo in May 2021 has been a timely reminder of the economic and environmental cost of a catastrophe at sea. A good example of the specialised ships required to police territorial seas is the French Naval Group's *L'Adroit* offshore patrol vessel (OPV) design. Four of these sophisticated but lightly-armed ships are now entering Argentine Navy service as its *Bouchard* class. Bruno Huriet's chapter provides an overview of the origin of these vessels and the much-needed surveillance capacity they provide to safeguard Argentina's exclusive economic zone (EEZ).

Another lightly-armed OPV design that is currently being delivered is the British Royal Navy's 'River' Batch II class. Ordered to maintain employment and skills in the River Clyde shipyards in advance of the delayed Type 26 Global Combat Ship programme, their arrival has fortuitously coincided with a new policy of the forward deployment of Royal Navy warships to maintain a national 'presence' across the world's seas. The editor's review describes the type's origins in more detail, as well as the capabilities – and limitations – they bring to this role.

An altogether different type of vessel is represented by Nikolai Novichkov's description of the Russian Navy's Project 20380 *Steregushchiy* class corvettes and their variants. Although only slightly larger than the Royal Navy OPVs, these are front-line warships that bristle with armament. Intended to secure Russia's littoral waters – including the Pacific Ocean 'bastion' for the country's strategic submarines – against high-end threats, they are proving to be a successful and numerous class that will play an important part in Russian green water naval operations. Also at home in a littoral environment are the US Navy's new ship-to-shore connectors, the replacement for the current Landing Craft Air Cushion (LCAC) hovercraft. Richard Scott's chapter describes how a US Navy-led, iterative design approach is seemingly meeting with more success than some of the more innovative but problematic American warship programmes of recent years.

TECHNOLOGICAL DEVELOPMENTS

Our final section on technological developments starts, as always, with David Hobbs' regular review of world naval aviation, this year including a review of the V-22 Osprey tilt-rotor family. Perhaps heralding the likely future drive for inter-operability between US Navy and allied assets, he also describes the significance of the integration of US Marine Corps F-35B strike fighters onboard *Queen Elizabeth* for the British carrier strike group's lengthy global deployment. One area where collaborative operations between allied navies are already firmly established is in the often-neglected field of mine countermeasures. Here, technological change is driving a transformation in operational concepts that looks set, over time, to replace the traditional mine countermeasures vessel with a 'toolbox' of largely unmanned and increasingly autonomous systems. Richard Scott assesses this change and the various components of the toolbox that the transformation relies on for its success. Meanwhile, Norman Friedman examines another frequently ignored area, assessing the importance of electro-optical devices as a supplement or alternative to radar in both surveillance and fire-control roles.

SUMMARY

As in many areas of life, the harsh impact of the ongoing COVID-19 pandemic has seemingly served to relegate the significance of world naval developments far into the background. However, the main themes of recent years – the relative growth of Asian navies on the back of shifting economic power and the rapid pace of technological change – have been little impacted. In similar fashion, the broader authoritarian challenge to the established democratic rules-based world order continues unabated from both within and without. The new administration in the White House clearly recognises the challenge, with its renewed emphasis on a collaborative international response a realistic recognition of the limitations of American power. Navies will have an important part to play in ensuring the success of this strategy and attempts to achieve greater levels of integration between 'Western' fleets look set to be an ongoing theme.

ACKNOWLEDGEMENTS

Seaforth World Naval Review continues to rely on the support of a small core team led by publisher Rob Gardiner, designer Steve Dent and proof reader Stephen Chumbley as it reaches its thirteenth year of publication. The technical illustrations provided by John Jordan and the enduring willingness of a range of distinguished authors to contribute their specialist expertise remain the other essential elements of the annual's ongoing success. The editor would also like to place on record his gratitude to Derek Fox, Tetsuya Kakitani, Marc Ottini, Marc Piché, Bernard Prézelin, Pete Sandeman, Chris Sattler, Guy Toremans, Xavier Vavasseur and Devrim Yaylali for the supply of images despite the continued constraints of the pandemic. Amongst many representatives from industry, the help of Babcock's Graeme Mair, BAE Systems' Kris Jones, Fincantieri's Cristiano Musella, Lürssen's Marei Winter and Navantia's Esther Benito Lope has been particularly appreciated. The customary assistance of my wife, Susan, in proof reading the initial drafts of my text remains evidence of the benefits that an enduring alliance can sustain.

The editor would also like to conclude by apologising for delays in the release of last year's edition. *Seaforth World Naval Review* experienced its own exposure to authoritarian trends with the refusal of the relevant censor to allow its printing in China, requiring a last-minute shift in this work to Singapore. This, combined with pandemic-related disruption, meant that the book's arrival was later than planned. Hopefully normal service will be resumed this year. In the meantime, all comments and criticism continue to be appreciated, please direct these for my attention to: info@seaforth-publishing.com.

Conrad Waters, Editor
30 June 2021

Notes

1. See *Renewing America's Advantages: Interim National Security Strategic Guidance* (Washington DC: The White House, 2021) available at: nssarchive.us/wp-content/uploads/2021/03/2021_Interim.pdf

2. See *Global Britain in a competitive age: The Integrated Review of Security, Defence, Development and Foreign Policy, Cm. 403* (London: HMSO, 2021).

3. Assessing how much the United Kingdom actually spends on defence has become increasingly complicated in recent years, partly because the public expenditure data published by the British Ministry of Defence has steadily diverged from the amount of spending reported to NATO. By mid-2019 this gap had grown to c. £8.5bn (£46.5bn compared with £38bn) or over 20 percent. SIPRI therefore undertook a detailed investigation of the discrepancy between the two figures, which ultimately resulted in their own calculations being adjusted upwards by around 10 percent. As SIPRI noted, this work '... shows that, even in a highly developed democracy (such as the UK), which releases numerous budgetary documents, it is not always possible to fully estimate total military spending.' See Nan Tian & Lucie Béraud-Sudreau, *Reassessing SIPRI's military expenditure estimate for the United Kingdom* posted to the SIPRI website on 9 February 2021 and available at: https://www.sipri.org/commentary/topical-backgrounder/2021/reassessing-sipris-military-expenditure-estimate-united-kingdom

4. To some extent, this decline was the result of a series of unforeseen events, including the destruction by fire of the amphibious assault ship *Bonhomme Richard* (LHD-6) and delays to acceptance of *Freedom* (LCS-1) Littoral Combat Ships due to design problems. Nevertheless, the US Navy still seems to be struggling to find both the design strategy and money to field the ships it needs in the numbers required.

5. Distributed Maritime Operations (DMO) is a concept under which technology is used to integrate the total warfighting potential of diverse, geographically dispersed ships, submarines and aircraft into a war-winning weapon. Manned-unmanned teaming (MUM-T) – the efficient coordination of manned and unmanned platforms – is likely to be one of the prerequisites for DMO to become a success.

2.1 REGIONAL REVIEW

Author:
Conrad Waters

NORTH AND SOUTH AMERICA

The advent of a new United States president has essentially confirmed that the large increases in military spending that were already starting to run out of steam as the Trump presidency came to an end are now definitively over. The FY2022 National Defense Budget Request released on 29 May 2021 amounted to US$752.9bn. This is a nominal 1.6 percent increase on the previous year's US$740.7bn but broadly flat overall in real terms. The Department of the Navy's allocation was US$211.7bn compared with last year's US$208bn, representing a similar level of growth. With some, more radical Democrat Members of Congress already seeking a more dramatic curtailment to military largesse, the navy will likely need to live within its existing funding envelope for the foreseeable future.[1]

What is less certain at the current time is how the US Navy will distribute the resources it has been allocated in an era when the challenges it faces are the greatest for many decades. Some choices seem to be clearer than others. The Global Posture Review that is currently underway has already been mandated to prioritise the growing threats from China and Russia over operations in more peripheral areas. It also seems likely from detailed budget breakdowns that fleet readiness and, particularly, longer-term research and development will be favoured over immediate procurement. These decisions seem to be rational given the large investments that will be required if the new networked and autonomous technologies that support operational concepts such as distributed maritime operations (DMO) are going to work effectively. DMO, alongside the US Marine Corps' new Littoral Operations in a Contested Environment and Expeditionary Advanced Base Operations concepts, was at the centre of the new tri-naval service *Advantage at Sea* strategy published in the dying days of the Trump administration in December 2020.[2]

There is greater doubt as to what structure the future US Navy should take. In December 2016, the navy released the results of a Force Structure Assessment (FSA) setting a goal for a 355-strong 'battleforce', a target that compares with fewer than 300 ships currently in service. Work has been undergoing since 2019 to update this target in view of the changing strategic situation and evolving technologies, particularly in the unmanned sphere. The general direction of travel has been a vision of a fleet with a smaller proportion of large ships, greater numbers of small ships and a new component of unmanned vessels to support the DMO strategy.

However, precisely how this analysis will translate into the fleet's constituent numbers remains under consideration. The latest update to Congress on long-range shipbuilding plans – the first issued by the Biden administration – is a holding document, setting broad force parameters that will not be fully refined until next year's budget cycle. In general terms, the new report seems to reflect budgetary realities by 'rowing back' on plans for expansion of the manned fleet whilst continuing to seek a large fleet of unmanned ships, albeit smaller than sometimes envisaged. However, the overall fleet is still targeted to be larger and, thus, more expensive than hitherto. This raises questions as to how such a navy is going to be built and maintained, particularly given the US Navy's long history of struggling to grow ship numbers.[3]

Another question is how to reverse the lacklustre results of recent US Navy warship procurement. The most significant surface combatant class – the *Arleigh Burke* (DDG-51) destroyers – trace their origins to the Cold War. Subsequent ships – the *Zumwalt* (DDG-1000) class destroyers and Littoral Combat Ships – have been design dead ends into which vast sums have been poured for modest tangible results. A shocking element of the latest budget is a proposal to retire the *Freedom* (LCS-1) variant ships *Detroit* (LCS-7) and *Little Rock* (LCS-9), fewer than five years after they entered service. A defect in the class's combining gearing that connects the ships' gas turbines and diesels with their waterjets has resulted in the navy refusing to accept further deliveries until the problem is resolved. It also appears it has determined that the cost of fixing the problem in *Detroit* and *Little Rock* is not worth the effort. Combined with the previous decision to retire the first four, prototype littoral combat ships due to a non-standard configuration, the navy will have written off six members of a class that was, until recently, supposed to be a mainstay of the future fleet. It can only be hoped that the US Navy can learn lessons from these past programmes as it stands up the program office for its next surface combatant, the DDG(X) guided missile destroyer.

The US Navy *Arleigh Burke* class destroyer *Barry* (DDG-52) conducts a replenishment exercise with the amphibious assault ship *America* (LHA-6) on 15 October 2020. The US Navy is struggling to determine the right mix of large ships such as these in its future force mix as it restructures to face ever-strengthening Chinese naval power. *(US Navy)*

Table 2.1.1: FLEET STRENGTHS IN THE AMERICAS – LARGER NAVIES (MID 2021)

COUNTRY	ARGENTINA	BRAZIL	CANADA	CHILE	COLOMBIA	ECUADOR	PERU	USA
Aircraft Carrier (CVN/CV)	–	–	–	–	–	–	–	11
Strategic Missile Submarine (SSBN)	–	–	–	–	–	–	–	14
Attack Submarine (SSN/SSGN)	–	–	–	–	–	–	–	54
Patrol Submarine (SSK)	–[1]	5	4	4	4	2	6	–
Fleet Escort (CG/DDG/FFG)	4	7	12	8	4	2	7	93
Patrol Escort/Corvette (FFG/FSG/FS)	9	2	–	–	2	6	2	24
Missile Armed Attack Craft (PGG/PTG)	2	–	–	3	–	3	6	–
Mine Countermeasures Vessel (MCMV)	–	3	12	–	–	–	–	8
Major Amphibious Units (LHD/LPD/LPH/LSD)	–	2	–	1	–	–	1	31

Note:

1. Argentina's two remaining submarines are non-operational and unlikely to be returned to service. A number of other Argentinian and Brazilian vessels are of uncertain operational status.

MAJOR NORTH AMERICAN NAVIES – CANADA

The most significant event to benefit the Royal Canadian Navy (RCN) in the past year has been the formal commissioning of the lead *Harry DeWolf* class Arctic and offshore patrol vessel at a ceremony at Halifax on 26 June 2021. The ship was delivered by Irving Shipbuilding's Halifax yard at the end of July 2020 and has subsequently completed an extensive series of post-acceptance trials and training. Her arrival in the operational fleet provides the navy with the first tangible fruits of the Canadian National Shipbuilding Strategy (NSS) announced as far back as 2010. Five other ships of the class will be completed for RCN service over the next few years followed by two slightly modified vessels for the Canadian Coast Guard. The second ship of the class, *Margaret Brook*, is currently undergoing trials whilst 2021 has also seen the keel laying for the fourth unit, *William Hall*, on 17 February as well as the first steel cutting for the fifth vessel, *Frédérick Rolette*, on 20 May.

The next project allocated to Irving's Halifax facility under the NSS is the new Canadian Surface Combatant. The design effort for the new ship is being led by Lockheed Martin Canada (LMC) and uses the BAE Systems Type 26 Global Combat Ship as its starting point. The last year has seen a steady flow of announcements of equipment orders for the first ships of the class. LMC's CMS 330 combat management system – used, inter alia, in the existing *Halifax* class frigates' modernisation – will interface with an Aegis module and the new AN/SPY-7 active phased array to control three layers of protective missile systems. These will encompass Standard missiles for area defence, Evolved Sea Sparrow missiles for point defence and Sea Ceptor missiles for last-ditch, close-in protection. Other equipment, such as the Ultra Electronics bow-mounted sonar and torpedo defence systems and the Rolls-Royce supplied propulsion plant, will be largely common with the Type 26. All-in-all, the Canadian Surface Combatant will be a large and very capable ship that has inevitably resulted in significant concerns over the ultimate bill. However, the Canadian government looks committed to the programme, with orders for a first batch of three ships from an eventual fifteen-strong frigate class expected in the course of 2023 prior to the start of construction during the following year.[4]

The new lead Royal Canadian Navy Arctic and Offshore Patrol Ship (AOPS) *Harry DeWolf* seen in the course of post-acceptance trials in company with the *Kingston* class coastal defence vessel *Summerside*. *Harry DeWolf* was formally commissioned on 26 June 2021. *(Canadian Armed Forces)*

The other shipbuilding facility heavily involved in RCN modernisation – Seaspan's Vancouver yard – continues to work on the lead ship of two *Protecteur* class joint support ships, which are based on the ThyssenKrupp Marine Systems (TKMS) Type 702 *Berlin* class design.[5] In March 2021 Seaspan reported that 90 percent of the vessel's constituent blocks were in production and that the major assemblies that formed the ship's bow section were in the course of being integrated. Delivery of *Protecteur* is scheduled for 2023, a deadline that will be closely watched given delays with earlier ships in the NSS programme. Her sister, *Preserver*, will be delivered later in the decade as the yard has been assigned work on a new Coast Guard oceanographic science vessel between the two ships. Also on the west coast, the *Victoria* class submarine *Corner Brook* will soon return to service for the first time since sustaining significant damage in a grounding in 2011. Her subsequent repair and refit – undertaken as part of a scheduled extended docking work period from 2014 onwards – has been delayed both by a fire and accidental damage to the boat's ballast system; the latest

Table 2.1.2: CANADIAN NAVY: PRINCIPAL UNITS AS AT MID 2021

TYPE	CLASS	NUMBER	TONNAGE	DIMENSIONS	PROPULSION	CREW	DATE
Principal Surface Escorts							
Frigate – FFG	**HALIFAX**	12	4,800 tons	134m x 16m x 5m	CODOG, 29 knots	225	1992
Submarines							
Submarine – SSK	**VICTORIA** (UPHOLDER)	4	2,500 tons	70m x 8m x 6m	Diesel-electric, 20+ knots	50	1990

in a series of mishaps to impact the class. Nevertheless, the completion of the work holds out the prospect of three of the four-boat class being available for operations in the near future; a note-worthy achievement given past problems.

Current fleet strength, unchanged year-to-year, is set out in Table 2.1.2.

MAJOR NORTH AMERICAN NAVIES – UNITED STATES

The US Navy had a total 'battleforce' of 297 warships as of mid-2021. In spite of the stated aim of increasing the fleet's size, this actually represented a small decrease year-on-year. The loss of the amphibious assault ship *Bonhomme Richard* (LHD-6) to fire exacerbated the impact of planned withdrawals whilst deliveries were lower than expected as no new *Freedom* (LCS-1) variant Littoral Combat Ships were accepted into service following the emergence of the combining gearing problems referenced above. The situation is not likely to improve significantly over the short term. The new administration looks set to continue plans to retire many of the remaining *Ticonderoga* (CG-47) class cruisers which – when coupled with premature retirements of Littoral Combat Ships – will counterbalance the beneficial impact of scheduled new ships.

At this moment, it is difficult to ascertain precisely what the longer-term future holds. The US Navy's 2016 FSA target of 355 manned warships was initially due to be replaced by a new plan informed by a study known as the Integrated Naval Force Structure Assessment (INFSA). This also incorporated US Marine Corps (USMC) structural requirements. However, publication of the INFSA's findings was apparently cancelled by the US Department of Defense around the end of 2019 due to discontent with the navy's conclusions. This resulted in INFSA being superseded by a new Future Naval Force Study (FNFS) led by the Office of the Secretary of Defense. Key conclusions from the FNFS were released as part of the *Report to Congress on the Annual Long-Range Plan for Construction of Naval Vessels* during the dying days of the Trump administration in December 2020. This also contained the structure of an indicative 2045 fleet sometimes referred to as the 'Battle Force 2045' plan. 'Battle Force 2045' has, in turn, been superseded by a far less ambitious series of new 'potential naval platform ranges' released by the new Biden administration in the *Report to Congress on the Annual Long-Range Plan for Construction of Naval Vessels for Fiscal Year 2022* released in June 2021. These will be refined into more specific plans over the next year. A summary of this complex and constantly evolving picture is provided in Table 2.1.3.

Table 2.1.3: UNITED STATES NAVY: RECENT FORCE STRUCTURE PLANS

TYPE	ACTUAL (MID-2021)	FSA 2016	FNFS RANGES	BATTLEFORCE 2045	INDICATIVE RANGES 06.2021
Aircraft Carriers	11	12	8–11	11	9–11
Light Carriers	–	–	0–6	–[4]	–[4]
Amphibious Assault Ships	9	12	9–10	9	8–9
Large Amphibious Ships	22	26	[52–57][3]	[57][3]	16–19
Small Amphibious Ships	–	–	[52–57][3]	[57][3]	24–35
Large Surface Combatants	92[1]	104	73–88	74	63–65
Small Surface Combatants	32[2]	52	60–67	66	40–45
Strategic Submarines	14	12	12	12	12
Attack Submarines	54	66	72–78	72	66–72
Combat Logistics Vessels	29	32	69–87	69	56–75
Support Vessels	34	39	27–30	33	27–29
Total Manned	**297**	**355**	**382–446**	**403**	**321–372**
Unmanned Surface	–	–	119–166	119	59–89
Unmanned Sub-Surface	–	–	24–76	24	18–51
Total Unmanned	**297**	**–**	**143–242**	**143**	**77–140**
Overall Total	297	355	525–688	546	398–512

Notes:

1. Excludes one *Zumwalt* (DDG–1000) class destroyer that has been commissioned but is not yet counted as part of the 'battle force'.
2. Includes eight MCMVs.
3. Split between large and small amphibious ships not stated.
4. Light aircraft carriers may be included in the force mix.

Taking a critical perspective, it is possible to argue that the vast variations in the numbers contained in these structure plans over such a short period of time suggests that the United States has not yet properly worked out what fleet it needs to respond to the emerging Chinse and Russian challenge. Whilst there may be some truth in such a criticism, some continuing themes are already evident. These include a likely decline in the relative importance assigned to manned surface combatants when considering both small and large vessels together; an increase in the value attached to submarines; and a very considerable future reliance on unmanned technologies. There will also have to be a substantial uplift in the logistical support force to sustain more widely dispersed units operating under a distributed maritime operations construct. The aircraft carrier looks set to remain an important part of the fleet, albeit with debate continuing over the merits of smaller carriers within the overall force structure.

Some additional observations seem pertinent. First, even the Biden administration's cut-back plans suggest a significant programme of naval expansion that will require large amounts of money at a time when building the new *Columbia* (SSBN-826) strategic submarines will put a severe strain on the naval budget. This suggests a tough fight for resources, potentially at the expense of other branches of the armed forces, if they are to be realised. The plans also involve a very meaningful shift in resources to new and untried technologies at the expense of tried and tested assets, as well as to the potential detriment of that part of the defence industrial base that sustains them. There will be a real struggle to achieve acceptance of this shift, a reality already demonstrated by pushback from Congress on some elements of the new plan.

For the time being, proposed procurement for

The US Navy's future structure plans place significant emphasis on unmanned and autonomous vessels despite some Congressional scepticism as to how well these new technologies will work. This April 2021 image shows the prototype autonomous catamarans *Seahawk* (front) and *Sea Hunter* departing harbour for the Unmanned Systems Integrated Battle Problem 2 manned-unmanned teaming exercise. *(US Navy)*

The new US navy aircraft carrier *Gerald R. Ford* (CVN-78) is dwarfed by a huge explosion during the first of a planned series of shock trials. The force of the blast – which took place 100 miles off the Florida coast – registered the equivalent of a 3.9 magnitude earthquake. *(US Navy)*

FY2022 is largely in line with the Future Years Defence Program (FYDP) previously set out in FY2021. It encompasses eight ships: two *Virginia* (SSN-774) class submarines; a single *Arleigh Burke* (DDG-51) destroyer; one *Constellation* (FFG-662) class frigate; one *John Lewis* (T-AO-205) fleet oiler, two TATS Towing and salvage ships and a TAGOS(X) surveillance vessel at a total cost of US$18.1bn. This represents one fewer destroyer but an additional oiler and another towing ship compared with the previous FYDP's eight-ship, US$17.8bn acquisition plan. The reduction in proposed destroyer construction – driven entirely by affordability – will impact a multi-year construction contract awarded to the class's builders and looks set to be resisted by Congress. No FYDP information has been provided due to the uncertainties outlined above.

Table 2.1.4 provides a snapshot of the current US Navy fleet, with a summary of recent developments relating to major warship types set out below.

Aircraft Carriers: The last year has seen the new carrier *Gerald R. Ford* (CVN-78) make further progress towards operational service, undertaking the first of a series of shock trials on Friday 18 June 2021. The test, which took place 100 miles off the Florida coast, registered the equivalent of a 3.9 magnitude earthquake. The last US Navy aircraft carrier to go through similar testing was *Theodore Roosevelt* (CVN-71) in the autumn of 1987. Completion of the trials will be followed by a period in dockyard hands for final defect rectification prior to the new carrier being cleared to commence work up for operational deployment. Future planned deliveries of *Ford* class carriers encompass *John F. Kennedy* (CVN-79) in 2024, *Enterprise* (CVN-80) in 2028 and *Doris Miller* (CVN-81) in 2032 under a procurement programme estimated to cost in excess of US$50bn.[5]

Surface Combatants: The US Navy continues to divide its surface combatants into large surface

Table 2.1.4: UNITED STATES NAVY: PRINCIPAL UNITS AS AT MID 2021

TYPE	CLASS	NUMBER	TONNAGE	DIMENSIONS	PROPULSION	CREW	DATE
Aircraft Carriers							
Aircraft Carrier – CVN	**FORD** (CVN-78)	1	100,000 tons+	333m x 41/78m x 12m	Nuclear, 30+ knots	4,600	2017
Aircraft Carrier – CVN	**NIMITZ** (CVN-68)	10	101,000 tons	333m x 41/78m x 12m	Nuclear, 30+ knots	5,200	1975
Principal Surface Escorts							
Cruiser – CG	**TICONDEROGA** (CG-47)	22	9,900 tons	173m x 17m x 7m	COGAG, 30+ knots	365	1983
Destroyer – DDG	**ZUMWALT** (DDG-1000)	2[1]	15,800 tons	186m x 25m x 8m	IEP, 30+ knots	175	2016
Destroyer – DDG	**ARLEIGH BURKE** (DDG-51) – Flight II-A	41	9,400 tons	155m x 20m x 7m	COGAG, 30 knots[1]	320	2000
Destroyer – DDG	**ARLEIGH BURKE** (DDG-51) – Flights I/II	28	8,900 tons	154m x 20m x 7m	COGAG, 30+ knots	305	1991
Littoral Combat Ship – FS	**FREEDOM** (LCS-1)	10	3,500 tons	115m x 17m x 4m	CODAG, 45+ knots	<50[2]	2008
Littoral Combat Ship – FS	**INDEPENDENCE** (LCS-2)	14	3,000 tons	127m x 32m x 5m	CODAG, 45+ knots	<50[2]	2010
Submarines							
Submarine – SSBN	**OHIO** (SSBN-726)	14	18,800 tons	171m x 13m x 12m	Nuclear, 20+ knots	155	1981
Submarine – SSGN	**OHIO** (SSGN-726)	4	18,800 tons	171m x 13m x 12m	Nuclear, 20+ knots	160	1981
Submarine – SSN	**VIRGINIA** (SSN-774)	19	8,000 tons	115m x 10m x 9m	Nuclear, 25+ knots	135	2004
Submarine – SSN	**SEAWOLF** (SSN-21)	3[3]	9,000 tons	108m x 12m x 11m	Nuclear, 25+ knots	140	1997
Submarine – SSN	**LOS ANGELES** (SSN-688)	28	7,000 tons	110m x 10m x 9m	Nuclear, 25+ knots	145	1976
Major Amphibious Units							
Amph. Assault Ship – LHD	**AMERICA** (LHA-6)	2	45,000 tons	257m x 32/42m x 9m	COGAG, 22+ knots	1,050	2014
Amph Assault Ship – LHD	**WASP** (LHD-1)	7[4]	41,000 tons	253m x 32/42m x 9m	Steam, 20+ knots	1,100	1989
Landing Platform Dock – LPD	**SAN ANTONIO** (LPD-17)	11	25,000 tons	209m x 32m x 7m	Diesel, 22+ knots	360	2005
Landing Ship Dock – LSD	**WHIDBEY ISLAND** (LSD-41)	11[5]	16,000 tons	186m x 26m x 6m	Diesel, 20 knots	420	1985

Notes:

1 Includes one ship commissioned but not fully operational pending completion of combat systems installation.

2 Plus mission-related crew.

3 Third of class, SSN-23 is longer and heavier.

4 LHD-8 has many differences.

5 Includes four LSD-49 HARPERS FERRY variants.

combatants – cruisers and destroyers – and small surface combatants – frigates, Littoral Combat Ships (LCS), mine countermeasures vessels (MCMVs) and patrol craft. Numbers of large surface combatants have been steadily increasing in recent years following resumed procurement of Flight IIA *Arleigh Burke* class destroyers from FY2010. The last year saw their numbers further increase by a further unit with delivery of the Bath Iron Works (BIW)-built *Daniel Inouye* (DDG-118) in March 2021. This trend is expected to accelerate in the coming years as BIW and Huntington Ingalls Industries (HII) work through a backlog of orders. From around 2023, deliveries will transition to the new, improved Flight III variant incorporating the new AN/SPY-6 Air and Missile Defence Radar (AMDR) starting with effect from *Jack H. Lucas* (DDG-125). The ship was launched by HII on 11 June 2021. The third and final *Zumwalt* class destroyer *Lyndon B. Johnson* (DDG-1002) should also be completed within the next two to three years. Debate continues as to how best these ships should be outfitted for their new surface strike role given the significant amount of bespoke technology they incorporate. However, it seems that previous thoughts around installing electromagnetic railguns in the class have now been superseded with plans to equip them with launchers for new hypersonic land attack missiles.

The flip side to these new arrivals is the acceleration of plans to withdraw the veteran *Ticonderoga* class cruisers, with a total of seven now slated for decommissioning over the next fiscal year. The navy is struggling to keep what are effectively life-expired hulls in an effective state against a backdrop of increasing repair and maintenance costs. The current aim seems to be to keep just enough of the class in commission to cover the key carrier strike group air defence command role the ships undertake during carrier deployments until they can be replaced by the new Flight III *Arleigh Burkes*. However, each cruiser carries 122 vertical launch system (VLS) missile cells; a loss that will be severely felt as the navy attempts to boost its strike capacity.[6]

Turning to smaller combatants, the last year has seen deliveries of two *Independence* (LCS-2) class variants – *Mobile* (LCS-26) and *Savannah* (LCS-28) – taking the number in service to fourteen. All but one of the remaining five members of the class are under construction following the start of work on *Kingsville* (LCS-36) at Austal's Mobile facility on 5 January 2021. However, the lead ship – *Independence* – will be withdrawn from service later in 2021 as part of previously announced plans to

Continued deliveries of *Arleigh Burke* class destroyers will be counterbalanced by planned withdrawals of many of the remaining *Ticonderoga* class cruisers, which are proving expensive to modernise and maintain. The HII-built *Delbert D. Black* (DDG-119) was officially handed over to the US Navy on 24 April 2020 but – partly because of the pandemic – it was not until 4 September that she finally left the building yard. *(Huntington Ingalls Industries)*

The *Freedom* variant Littoral Combat Ship *St. Louis* (LCS-19) pictured transiting the St. Lawrence Seaway in June 2020. The emergence of a design problem with the class's gearing has meant that no further ships of the class have been accepted into service. *(Marc Piché)*

dispose of the first pair of prototype ships of each LCS variant due to systems' incompatibilities with later members of the class. This plan will also see *Freedom* decommissioned over the coming months. No further members of this variant have joined the fleet since the delivery of the tenth ship – *St Louis* (LCS-19) – in June 2020 due to the need to resolve the gear train defects previously discussed. The Fincantieri Marinette Marine yard that builds the type currently has a backlog of six members of the class at various stages of construction. It will then progress to building the new *Constellation* class frigates. The second of these ships – *Congress* (FFG-63) – was ordered on 20 May 2021.

In spite of the delays to the LCS programme – and its associated mine warfare modules – the navy is pushing ahead with plans to withdraw its existing MCMVs. *Champion* (MCM-4), *Scout* (MCM-8) and *Ardent* (MCM-12) were all decommissioned in August 2020. This leaves eight of the original fourteen-strong *Avenger* (MCM-1) class in service. Somewhat below the radar screen, the navy has also been accelerating disposal of small combatants that do not appear in the official 'battle force' list, including the three 400-ton *Cyclone* (PC-1) class patrol boats assigned to home waters.[7] They will be followed by the twelve relatively new 70-ton Mk VI fast patrol boats delivered by SAFE Boats International between 2016 and 2018 for riverine and littoral operations. The current thinking appears to utilise Coast Guard assets to replace the gap created by these withdrawals. It is believed around US$200m has been spent acquiring and supporting the Mk VI boats in recent years; an investment that will now effectively be written off in another example of muddled US Navy procurement planning.[8]

Acquisition of the new generation of unmanned surface vessels that will supplement the manned fleet is still in the embryonic stage. The navy eventually hopes to field both missile-armed large unmanned surface vehicles (LUSVs) and sensor-carrying medium unmanned surface vehicles (MUSVs) but is being forced to move ahead slowly in the face of Congressional scepticism. In July 2020 a team led by L3Harris was awarded a US$35m contract for the design and construction of a MUSV prototype that could eventually result in the exercise of options for a further eight vessels. Subsequently, in September 2020, six contracts worth US$7m each were agreed with Austal USA, Bollinger Shipyards, Gibbs & Cox, HII, Lockheed Martin & Marinette Marine to produce design concepts for the LSUV. This should lead to the procurement of a prototype vessel during FY2022 and deployable ships from FY2023 onwards. In the meantime, testing of existing unmanned concept vessels in conjunction with manned warships is being stepped up by the US Navy's Surface Development Squadron 1, which was stood up at San Diego in May 2019. The squadron is slowly expanding, most recently taking delivery of a second autonomous *Sea Hunter* type catamaran, *Seahawk*, in April 2021. Later that month, the two ships played an important role alongside manned squadron member *Michael Monsoor* (DDG-1001) and other unmanned and manned assets in Unmanned Systems Integrated Battle Problem 2, an early trial of the manned-unmanned teaming approach that will be fundamental to the DMO concept.

Amphibious and Support Shipping: The US Navy's amphibious fleet suffered a major loss in July 2020 when the newly-refitted amphibious assault ship *Bonhomme Richard* was destroyed by fire. The conflagration broke out after an explosion whilst the

A team led by L3Harris has been awarded a contract for the design and construction of a prototype of the sensor-carrying medium unmanned surface vehicles (MUSVs) that are destined to play a major part in future US Navy operations. This is a graphic of the ship's design. *(L3Harris)*

The amphibious assault ship *Bonhomme Richard* (LHD-6) seen during the evening of 12 July 2020, several hours after a blaze broke out aboard the vessel. It took some four days to extinguish the fire and the damage was so great that it was decided to decommission and scrap the newly-refitted ship. *(US Navy)*

The first *John Lewis* (T-AO-205) class fleet replenishment oiler was launched from the General Dynamics NASSCO yard on 12 January 2021. This photograph was taken shortly before the ship's launch. *(General Dynamics NASSCO)*

ship was alongside at San Diego in the morning of 12 July 2020 and had damaged eleven of the ship's fourteen deck levels by the time it was extinguished some four days later. Analysis of the cause of the blaze and its rapid spread remain under investigation amidst speculation that arson might have been involved. However, the fact the ship had still to be fully restored to operational condition after her recent upgrade was likely an exacerbating factor. The cost of repairing the 22-year-old ship – some US$2.5bn to US$3.2bn – was deemed to be prohibitively expensive. Accordingly, she departed on her final voyage to the scrapyard in mid-April 2021.

There have been no deliveries of major amphibious vessels in the last year, although *Tripoli* (LHA-7) was commissioned on 15 July 2020 after having been delivered the previous February. Conversely, a start has been made on withdrawing the *Whidbey Island* (LSD-41) dock landing ships with the departure of *Fort McHenry* (LSD-43) from the fleet in March 2021. Whilst the initial Batch 2 *San Antonio* (LPD-17) class amphibious transport docks that are intended to replace the LSD-41 type have already been ordered, this new class could well be truncated if force structure plans that suggest fewer large amphibious ships will be required in future hold good. Their place will be taken by larger numbers of the new light amphibious warship (LAW), which are intended to support the USMC's new Expeditionary Advanced Base Operations (EABO) strategy. These will seemingly fall somewhere between the traditional large amphibious ships and auxiliary vessels like the *Spearhead* (T-EPF-1) expeditionary fast transports in terms of size and capability. Twelve of the latter are now in service following delivery of *Newport* (T-EPF-12) in September 2020. Another three have been ordered following the award of a contract for the future T-EPF-15 in February 2021. There had previously been some doubt over her future following the Trump administration's decision to shift funds authorised for her build to help pay for the Mexican border wall. However, Congress added the necessary money to FY2021's defence budget authorisation to ensure her construction.

January 2021 saw the launch from General Dynamics NASSCO of *John Lewis* (T-AO-205), the lead ship of a new class of fleet replenishment oiler. Six of the ships have been ordered to date and a seventh is included in the FY2022 budget request. The 2016 FSA envisaged a requirement for twenty of the ships but, as for so much of the US Navy's shipbuilding programme, this is up in the air pending the Biden administration's decisions on future fleet structure. What does seem likely is that the *John Lewis* class will increasingly be supplemented by a new next-generation logistic ship (NGLS). In similar fashion to the proposed LAW, this will be a smaller and cheaper vessel that will be built in the numbers required to support a more distributed force.

Submarines: The need to stabilise and then rebuild the number of nuclear-powered attack submarines whilst embarking on production of the next-generation *Columbia* class strategic submarines is one of the US Navy's greatest challenges. The main problem is that the industrial base required to ramp up production was severely eroded in the aftermath of the Cold War, with production limited to just one attack submarine each year as late as FY2010. Future plans envisage orders for at least two attack submarines and one strategic submarine p.a. from the middle of the decade; a major expansion.

Submarine manufacturing yards General Dynamics Electric Boat at Groton CT and HII Newport News are doing what they can to rise to the challenge. However, the overall result is proving to be something of a mixed success. On the downside, a previously enviable record of producing *Virginia* class attack submarines on time and to budget is starting to slip, with delivery dates continuously exceeding scheduled deadlines that have been made significantly shorter than when construction of the class first began. The impact of the COVID-19 pandemic has had a further negative impact and no *Virginia*s were actually delivered in the last twelve months. This resulted in the submarine force falling by two boats as at mid-2021 due to continued retirements of the life-expired members of the *Los Angeles* (SSN-688) class. The navy is planning to extend the lives of six of these boats – including a refuelling of their nuclear reactors – to squeeze, perhaps, an additional ten years of service out of each hull as part of efforts to retain overall submarine numbers around existing levels. *Cheyenne* (SSN-773) will be the first of the class to embark on this programme at the start of 2022.

In spite of production delays, it does seem that the submarine manufacturers have broadly managed to achieve the step-up in *Virginia* class production required, with a drumbeat of two attack submarine orders each year now firmly established. This was reinforced by Congress's decision to add a second boat to the FY2021 naval procurement budget,

Using a transfer car system, the *Virginia* class submarine *Montana* (SSN-794) is seen being transferred from her construction hall to a floating dry dock at HII's Newport News Shipbuilding yard on 15 October 2020 in preparation for her launch. US Navy submarine construction is struggling to keep to schedule as production demands continue to ramp up. *(Huntington Ingalls Industries)*

overturning the administration's somewhat surprising decision to request just one attack submarine for that year. In April 2021, a US$2.4bn contract modification to build the additional submarine was made to the c. US$22bn multi-year agreement that already covered nine 'Block V' boats previously awarded to the General Dynamics-HII consortium responsible for the class's production. Like most of the rest of this batch of submarines, the new boat will incorporate the Virginia Payload Module that adds additional firepower to the class.

Total *Virginia* class orders now extend to thirty-eight units (of which thirty-six had been authorised through to FY2022) and it seems that further members of the class will be approved throughout the rest of the decade. However, attention is already starting to turn to the next generation SSN(X). In contrast to the *Virginia* class, which was designed for multi-mission dominance in the littoral, SSN(X) will be designed for greater transit speed under increased stealth conditions in all ocean environments, and carry a larger inventory of weapons and diverse payloads. It will also be equipped to counter the threat posed by unmanned underwater vehicles (UUVs) and be able to coordinate a larger contingent of off-hull vehicles, sensors and friendly forces.[9] The US Navy awaits delivery of its own prototype 'Orca' extra-large unmanned undersea vehicles (XLUUVs), five of which were ordered from Boeing in 2019.

Meanwhile, a major step forward with *Columbia* strategic submarine class production was taken in November 2020 when General Dynamics Electric Boat was awarded a US$9.5bn contract for the detailed design and construction of the first two members of the twelve-boat class. Work is already underway on the four constituent models of the lead submarine allocated to Electric Boat – another two are assigned to HII – at the company's Rhode Island complex. They will ultimately be transferred by barge to a final test and assembly yard in Groton, where the components will be assembled into a complete submarine in a 200,000ft^2 facility currently under construction. Delivery is scheduled by the end of 2027, a tight deadline that will doubtless cause both navy and contractor many sleepless nights in the years ahead. The second submarine – named *Wisconsin* (SSBN-827) – will follow around three years later, with subsequent boats scheduled to be delivered at annual intervals thereafter. Consistently described as the US Navy's 'top priority program', estimated completion of the twelve boats remains set at c. US$10bn.

Operational Highlights: The 'operational context' overview provided in the press briefing that accompanied the US Navy's FY2022 budget request suggests that the fleet remains heavily committed, with 110 ships of the then 294 ship total expected to be deployed as at the end of FY2021. The Pacific Ocean continues to be the top area of operational focus, with 51 ships operating in the region and a further 110 based on the Pacific coast. Efforts are already underway to start testing out the new operational doctrines that would be employed in any new Pacific War, including the evolving USMC's EABO concept. This envisages small, highly-mobile and dispersed combat units penetrating an opponent's outer defences to land on the numerous islands and atolls that proliferate across the region. These would be used as bases to conduct quick, coup-de-main type strikes on enemy forces before rapid relocation to avoid counter-attack. A major early test of this scenario was conducted during Exercise Noble Fury in October 2020, when a unit from the US 2nd Marine Regiment simulated the capture of an airfield on the Japanese island of Ie Shima. This was then used to fly in a rocket artillery battery which performed an anti-shipping strike mission before immediately departing to another location. A further series of exercises – expanded to include regional allies – is planned for the coming year.[10]

A key element of US Navy activity in recent years has been an expansion in the freedom of navigation operations – commonly abbreviated as FONOPS – aimed largely at countering the maritime claims of revisionist powers. Those in the South China Sea have garnered most attention but recent deployments in waters neighbouring Russia – whilst not a formal part of the FONOPS programme – can be viewed as forming part of the same strategy. Interestingly, the FONOPS programme has also been aimed at perceived 'excessive claims' by a number of allied or friendly nations, including Brazil and Japan. In April 2021, this resulted in a minor spat with India when the destroyer *John Paul*

Jones (DDG-53) sailed west of Lakshadweep Islands, which are within India's exclusive economic zone.[11]

Although the United States is attempting to de-emphasise its footprint in the Middle East, tensions between the US Navy and Iranian forces in the region remain high. A number of incidents have occurred during which Islamic Revolutionary Guard Corps (IRGC) boats have harassed US Navy and US Coast Guard vessels operating in the Persian Gulf, resulting in the firing of warning shots on more than one occasion. There is a clear danger that any reduced American presence resulting from the Global Posture Review will encourage such action with potentially dangerous consequences. The refocusing of US Navy forces is also making its presence in South America. Here the USMC's Special Purpose Marine Air-Ground Task Force–Southern Command – used to support local allies in training, internal security and humanitarian operations – is being shut down as the Corps looks to concentrate on its new Pacific mission. It has to be questioned whether the saving – reportedly just US$3m – is worth the likely loss of influence in the United States' own back yard.

US Coast Guard: The US Coast Guard's fleet renewal plan has continued to make progress over the past year. The plan, which dates back as far as 2004, envisages the replacement of existing ships with a fleet of eight large oceanic national security cutters, twenty-five intermediate sized offshore patrol cutters and fifty-eight coastal fast response cutters.

The large ships are the 'Legend' or *Bertholf* (WMSL-750) class. Due to Congressional largesse, eleven of this type have actually been ordered to date and nine are now in service following the commissioning of *Stone* (WMSL-758) on 19 March 2021. HII's Ingalls division has built all the class, commencing fabrication of the final vessel – *Friedman* (WMSL-761) – in May 2021. They replace the twelve 1960s-era *Hamilton* (WHEC-715) high endurance cutters, which have all now been withdrawn from service following the decommissioning of the final member of the class, *Douglas Munro* (WHEC-724), on 24 April 2021.

The only slightly smaller and less capable 'Legend' class offshore patrol cutters will replace twenty-nine medium endurance cutters. Construction of the four initial members of the class has been allocated to Eastern Shipbuilding in Panama City Florida and the final vessel of this batch is included within the US Coast Guard's FY2022 budget request. The company was originally due to build more of the ships but an open competition is currently underway for the next eleven vessels after the shipyard ran into difficulties following damage from Hurricane Michael in December 2018. Meanwhile Congress funded the final four 'Sentinel' class fast response cutters in the FY2021 budget, taking total authorisations to the fifty-eight vessels in the renewal plan plus an additional six for service in the Persian Gulf. The craft are being built to Damen's Stan Patrol 4708 design by Bollinger Shipyards, which had delivered forty-five of the type by mid-2021.

The other major Coast Guard procurement programme is for new Polar Security Cutters. Two of a planned three heavy icebreakers have been funded to date and should be followed by three medium icebreakers in due course. The first of the new vessels should start construction at VT Halter Marine before the end of 2021 for delivery in 2024. The urgent requirement for the new ships was demonstrated by a propulsion train fire aboard the existing medium icebreaker *Healy* (WAGB-20) – one of only two operational Coast Guard icebreakers – in August 2020, severely limiting the service's ability to conduct polar operations until complex repairs could be effected.

A F-35B Lightning II strike fighter from Marine Fighter Attack Squadron 121 (VMFA-121) the 'Green Knights' refuels at an established Forward Arming and Refuelling Point during simulated Expeditionary Advanced Base Operations (EABO) in March 2019. The new concept is seen as an essential part of any future warfighting in the Pacific. *(US Marine Corps)*

OTHER NORTH AND CENTRAL AMERICAN NAVIES

Although **Mexico's** navy remains the only maritime force of note in the rest of North and Central America its modernisation has essentially ground to a halt under the current Obrador administration. The existing 2019–24 procurement programme envisages the acquisition of as many as thirty-six new vessels for the navy in Mexican yards. These include two additional 'POLA' (Damen 'Sigma' 10514) corvette-like oceanic patrol vessels and four *Tenochtitlan* class (Damen Stan Patrol 4207) inshore patrol vessels. However, funding has not been provided to progress construction of anything more than six small seaweed collection boats against a backdrop of official disapproval of the navy's partnership with foreign private enterprise to progress new projects. The only bright spot has been the delivery of the eighth and final *Oaxaca* class OPV. The new *Tabasco* was commissioned on 1 June 2021.[12]

Turning to Central America and the Caribbean, the most significant news is possibly the pending arrival of two new 58m Austal 'Cape' class patrol boats for the **Trinidad and Tobago** Coast Guard. The new *Port of Spain* and *Scarborough* were handed over at Austal's Henderson facility in May 2021 prior to departure for the long voyage to their new home aboard a heavy lift ship in June. Other new arrivals include the first examples of a series of 'Near Coastal Patrol Vessels' being supplied to countries across the region as part of a US Department of Defense initiative to strengthen regional security. The first of these vessels – Metal Shark 85 'Defiant' type boats based on Damen's Stan Patrol 2606 hull form – was incorporated into the **Dominican Republic's** navy as *Betelgeuse* at the Las Calderas Navy Base on 6 August 2020. Up to thirteen are scheduled for delivery to partner nations in the region, with **El Salvador**, **Guatemala** and **Honduras** amongst other nations due to receive examples of the type. **Costa Rica** has also been receiving Metal Shark boats, with smaller 'Defiant 45' and 'Defiant 38' craft reportedly delivered at the end of 2020.

MAJOR SOUTH AMERICAN NAVIES - BRAZIL

Brazil is currently the only Latin American country with active programmes for the construction of front-line naval combatants.[13] Major projects for the delivery of new submarines and frigates are in progress but both are seemingly being impacted by the severity of the COVID-19 pandemic that continues to hold Brazil in its grip. Reports suggest that this is affecting both the ability of industry to maintain design and construction activity and also the funding required to pay for the programmes that are underway.

The giant PROSUB construction project has made some headway in the last year. In July 2020, the Brazilian Navy formally inaugurated its new submarine base at Itaguaí – part of a wider submarine construction and maintenance complex – that will be home for the four 'Scorpène' type boats being completed under the programme. However, the lead submarine *Riachuelo*'s programme of tests and sea trials could well have been impacted by an accident in March 2021 that caused salt water flooding of her bow area. Current reports suggest that her delivery is now scheduled for the end of 2021, a further shift to the right. Her sister, *Humaitá*, was launched on 11 December 2020 and the hull of *Tonelero*, the

The first Brazilian 'Scorpène' type submarine *Riachuelo* has spent much of the last year undergoing trials from the giant submarine manufacturing complex and operating base at Itaguaí. She is expected to commission around the end of 2021 and will subsequently be followed by three further boats. *(Naval Group)*

Table 2.1.5: BRAZILIAN NAVY: PRINCIPAL UNITS AS AT MID 2021

TYPE	CLASS	NUMBER	TONNAGE	DIMENSIONS	PROPULSION	CREW	DATE
Principal Surface Escorts							
Frigate – FFG	**GREENHALGH** (Batch I Type 22)	2	4,700 tons	131m x 15m x 4m	COGOG, 30 knots	270	1979
Frigate – FFG	**NITERÓI**	5	3,700 tons	129m x 14m x 4m	CODOG, 30 knots	220	1976
Corvette – FSG	**BARROSO**	1	2,400 tons	103m x 11m x 4m	CODOG, 30 knots	145	2008
Corvette – FSG	**INHAÚMA**	1	2,100 tons	96m x 11m x 4m	CODOG, 27 knots	120	1989
Submarines							
Submarine – SSK	**TIKUNA** (Type 209 – modified)	1	1,600 tons	62m x 6m x 6m	Diesel-electric, 22 knots	40	2005
Submarine – SSK	**TUPI** (Type 209)	4	1,500 tons	61m x 6m x 6m	Diesel-electric, 22+ knots	30	1989
Major Amphibious Units							
Helicopter Carrier – LPH	**ATLÂNTICO** (OCEAN)	1	22,500 tons	203m x 35m x 7m	Diesel, 18 knots	490	1998
Landing Ship Dock – LSD	**BAHIA** (FOUDRE)	1	12,000 tons	168m x 24m x 5m	Diesel, 20 knots	160	1998

third of the quartet, closed up on the same day. Their deliveries are now scheduled to take place early in 2023 and 2025, with the final boat – *Angostura* – bringing up the rear in 2025. The preliminary design of the planned follow-on nuclear attack submarine *Alvaro Alberto* was reportedly completed at the end of 2020, paving the way for the award of initial construction contracts. However, even if all goes well – by no means a certainty given previous delays – it will not be before the mid-2030s that she becomes operational.

The vast sums being allocated to submarine construction are squeezing other naval programmes. However, conclusion of contracts early in 2020 with the TKMS/Embraer-controlled Águas Azuis consortium for a quartet of new *Tamandaré* class MEKO type frigates does hold out the prospect of at least partial renewal of Brazil's aging surface fleet. A number of contracts for long-lead items for the ships have subsequently been confirmed over the past year but actual construction of the first ship is not expected to commence before 2022. This suggests anticipated delivery in 2025 – already pushed back a year – may prove optimistic. The only other significant warship deliveries expected before then are the two final *Macaé* class OPVs *Maracanã* and *Mangaratiba*, the third and fourth members of what was initially planned to be a numerous type. The ships are being completed at Rio de Janeiro Navy Arsenal following the financial collapse of the EISA shipyard initially tasked with their construction.

Current Brazilian fleet strength – unchanged year-on-year – is set out in Table 2.1.5. Away from the headline numbers, the major departure was the MCMV *Albardão*, which was decommissioned in May 2021. This leaves just three minehunters in operational service.

Chile's ASMAR is expected to commence construction of four new amphibious and military transport vessels based on a VARD 7 series concept once work on a new icebreaker has been completed. *(VARD)*

OTHER SOUTH AMERICAN NAVIES

Elsewhere in South America, current projects are largely focused on improving constabulary and maritime logistical support capabilities. Although a number of the region's navies have ambitions for replacement frigate and submarine construction – the former often having a local element – these are unlikely to gain traction in the current environment considering the ongoing social and economic devastation being caused by the pandemic.[14]

Chile continues to have the region's most up-to-date and effective combat force. The last year has seen it continue modernising its existing trio of Type 23 frigates on the basis of a project built around LMC's CMS 330 combat management system. *Almirante Lynch* (the former HMS *Grafton*) has now finished the upgrade at ASMAR's Talcahuano yard, leaving only *Almirante Condell* (formerly *Marlborough*) to complete the programme. The navy envisages maintaining a fleet of eight surface combatants and might look to acquire one of the additional British Royal Navy Type 23 frigates slated for withdrawal under the latest defence review to replace its elderly Type 22 frigate *Almirante Williams* (previously *Sheffield*). The two FFG-7 class frigates acquired from Australia in 2019 and a pair of former Dutch 'M' class frigates make up the rest of the flotilla.

The longer-term aim is to build new frigates at ASMAR but the company is currently fully occupied with completing the 'VARD 9-203' type icebreaker *Antártica 1*, now scheduled for delivery in 2023. It will then commence work on a quartet of LSD-like amphibious and military transport vessels, which will be based on a VARD 7 series concept under a contract awarded in February 2021. These will eventually replace the two existing 'BATRAL' type tank landing ships, the transport *Aquiles* and the amphibious transport dock *Sargento Aldea* (the former French *Foudre*). The navy also needs to replace the elder pair – comprising two Type 209/1400 Simpson class boats – of its four-strong underwater flotilla but there has been no recent news of developments in this regard.

In spite of the first arrivals of the new *Bouchard* class OPVs, **Argentina's** navy remains in rather desperate straits. Its two remaining submarines are inoperable and – with the apparent cancellation of the refit of the more modern *Santa Cruz* – it seems unlikely that either will ever go to sea again. In March 2021, a number of former high ranking Argentine officers were subject to sentences varying from close arrest to dismissal from the service for acts of omission with respect to the loss of her sister, *San Juan*, in November 2017. Meanwhile immediate naval priorities appear to encompass completion of two 36m cadet training ships by the local Astillero Rio Santiago yard as a precursor to other indigenous construction.

Colombia's COCTEMAR and **Peru's** SIMA both continue to build new second-line but effective ships to their respective navies. The former delivered the sixth 'BDA/BAL-C' logistic support landing ship *Bahía Solano* in March 2021 and hopes to receive orders for new OPVs to its improved OPV-93C design under the Naval Development Plan 2042 that was published in mid-2020. The Colombian Navy was also bolstered by the delivery of another former South Korean corvette – the *Po Hang* type *Almirante Tono* (previously *Iksan*). SIMA completed the fifth and sixth *Río Pativilca* coastal patrol ships *Río Tumbes* and *Río Locumba* early in 2021 and is scheduled to be involved in a new OPV construction programme for up to three vessels once the Callao shipyard has finished work on the second *Paita* (*Makassar*) class amphibious transport dock *Pisco.*

Other South American countries with active patrol vessel programmes include **Ecuador** and **Uruguay.** The former received two former South Korean Coast Guard *Hae Uri* class coastal patrol vessels at the end of 2020 and its own local shipbuilding group is building what could be the first of a series of German Fassmer MPV70 Mk III multirole vessels. For its part, Uruguay has negotiated the transfer of three comparatively modern 'Marine Defender' (Stan Patrol 2600 series) coastal patrol vessels from the US Coast Guard as it continues to seek to progress its longstanding requirement for up-to-date offshore patrol vessels.

Notes

1. These figures are taken from the budget briefing materials released by the Office of the Undersecretary of Defense (Comptroller) and available at: comptroller.defense.gov/Budget-Materials/. In inflation adjusted, constant FY2010 terms the FY2022 naval budget request was US$170.3bn compared with US$171.6bn last year, a recent high of US$177.5bn in FY2020 and US$177.1bn in 2010.

2. See: *Advantage at Sea – Prevailing with Integrated All-Domain Naval Power* published jointly by the US Navy, the US Marine Corps and the US Coast Guard in December 2020.

3. Further analysis on the recent development of the US Navy's future force structure is contained later in this chapter. A more detailed assessment can be found in Ronald O' Rourke's *Navy Force Structure and Shipbuilding Plans: Background and Issues for Congress RL32665* (Washington DC: Congressional Research Service, 2021). Like many of Mr O'Rourke's other periodically updated reports, this has been posted as a public service on the Federation of American Scientists' website at: fas.org/sgp/crs/weapons/

4. See David Pugliese's 'Government to commit to building first three warships despite budget concerns' posted to the *Ottawa Citizen* website – ottawacitizen.com – on 8 June 2021.

5. See Ronald O' Rourke, *Navy Ford (CVN-78) Class Aircraft Carrier Program: Background and Issues for Congress RS2064* (Washington DC: Congressional Research Service, 2021), p.10.

6. An excellent overview of the different considerations impacting withdrawals of the *Ticonderoga* class was provided by Megan Eckstein, 'Navy "Struggling" to Modernize Aging Cruiser Fleet As Tight Budgets Push Pentagon to Shed Legacy Platforms' posted to the *USNI News* site – news.usni.org on 5 April 2021.

7. A further ten members of the *Cyclone* class currently operate in the Persian Gulf. They are also drawing towards the end of their service life.

8. A strong critique of the decision to retire the Mk VI patrol boats was provided by Christopher P Cavas in 'A Problem of Commitment' posted to the *Defense & Aerospace Report* site – defaeroreport.com – on 11 April 2021. Comparing the proposed light amphibious warship (LAW) to the Mk VI debacle, he wrote: 'The LAW, in other words, is the next shiny object, the latest infatuation, in a line of ship types leading back in time from the Mark VIs to the littoral combat ship to the patrol coastals ... But if repeated examples of recent history are any guide, the next generation of Navy and Marine officers and planners with their own priorities and viewpoints are very likely to urge the craft be discarded, probably even before the initial group is in service.'

9. For additional information on the evolving SSN(X) development effort see another Ronald O'Rourke report, *Navy Next-Generation Attack Submarine (SSN[X]) Program: Background and Issues for Congress IFF 11826* (Washington DC: Congressional Research Service, 2021).

10. For more information on the USMC's efforts to adapt to the EABO concept see Sidney E. Dean, 'A Renaissance in Amphibious Warfare', *Maritime Security & Defence* – June 2021 (Bonn: Mittler Report, 2021), pp.53–7.

11. The US Department of Defense issues an *Annual Freedom of Navigation Report*, the latest located at: policy.defense.gov/Portals/11/Documents/FY20%20DoD%20FON%20Report%20FINAL.pdf. Indian ire at the *John Paul Jones* transit was reported in an article by Rezaul H. Laskar and Rahul Singh, 'India objects to US Navy operations near Lakshadweep' posted to the *Hindustan Times*' website – hindustantimes.com – on 10 April 2021.

12. Interestingly, the Mexican Navy's responsibilities have been increased significantly over the last year in spite of its procurement woes as it has been granted responsibility for managing the country's commercial ports and mercantile marine department. The aim, apparently, is to root out longstanding corruption.

13. The *Poder Naval* site – www.naval.com.br/blog/ – remains an essential source for detailed information on Brazilian Navy developments.

14. Along with official navy websites, the Spanish-language infodefensa.com and its rival defensa.com remain invaluable sources of information on naval developments in Central and South America. A good overview of the main regional navies was also provided by Tim Fish in 'Briefing: South American naval fleets: keeping up appearances', *Jane's Defence Weekly* – 18 November 2020 (Coulsdon: Jane's, 2021), pp.22–7

2.2 REGIONAL REVIEW

ASIA AND THE PACIFIC

Author:
Conrad Waters

Amongst the world's larger fleets, it is undoubtedly the Republic of Korea Navy that attracts the least attention. South Korea has long had the economic means and technological capacity to become a major sea power. Moreover, the country's status as a global trading nation has provided it with every incentive to build the capability to protect the maritime trade routes on which the nation's economic prosperity depends. However, the obvious impediment to the creation of an oceanic fleet has been the ever-present threat from the unstable North. This has prioritised the importance of the army whilst emphasising the significance of the navy's littoral role as its key area of focus.

Despite this constraint, the Republic of Korea has long held 'blue water' ambitions. It has been working steadily – and with considerable success – to achieve its objectives. Several factors have been working in its favour. One has been the reduction of army dominance of political decision-making as the country has democratised. This has made the acceptance of a national maritime vision a more attainable aim than hitherto. More recently, development of strategic concepts in the light of North Korea's ballistic missile and nuclear warhead programmes have seen emphasis placed on (1) a Korean Air and Missile Defense (KAMD) System and (2) a Strategic Strike System as key elements of deterrence.[1] These concepts play to the strengths of sophisticated naval assets such as Aegis-equipped destroyers and large missile-carrying submarines that are also key elements of oceanic naval operations. Moreover, growing naval rivalry amongst South Korea's neighbours has not passed unnoticed amongst the country's political leadership.

The growing attractions of naval power have been accompanied by a well thought-out and persistent industrial plan. This has seen South Korea gradually acquire the means to undertake a broad range of naval construction activity with only limited recourse to foreign partners. This strategy has evolved from the local construction of vessels with overseas design assistance through the production of local designs with significant levels of foreign equipment to ships that are now largely designed and outfitted by indigenous companies. In the course of this journey, South Korean vessels have become increasingly attractive on international export markets. This is evidenced, for example, by recent sales of frigates to the Philippines and Thailand.[2]

The last year has seen a further stage in the Republic of Korea Navy's rise to the status of an international naval power. The announcement of the South Korean Ministry of Defence's Intermediate-Term Defence Plan for the period 2021–5 in August 2020 – encompassing a total 35 percent increase in spending over its five-year lifespan – included confirmation that previous proposals for a new LPX-II amphibious assault ship had morphed into a full-scale aircraft carrier. Now referred to as the CVX, the new aircraft carrier is targeted to have a light displacement of c. 30,000 tons. However, competing conceptual designs from Hyundai Heavy Industries (HHI) and Daewoo Shipbuilding & Marine Engineering (DSME) displayed at the South Korean MADEX maritime exhibition in June 2021 revealed ships with lengths in the region of 260–270m and a full load displacement of 45,000 tons or more. Both are fitted-out with twin islands and will be optimised for F-35B Lightning II short take-off and vertical landing (STOVL) operation. The United Kingdom's Babcock and Italy's Fincantieri – respectively builders of the *Queen Elizabeth* class aircraft carriers and *Trieste* amphibious assault ship – have been brought in to help refine the competing concepts prior to selection of one or other to move forward to detailed design by the end of 2022. If considerable local opposition to the programme – largely on grounds of cost and utility – can be overcome, it is hoped that the new vessel will enter service in the first half of the 2030s.

At the other end of the capability scale, a major milestone in Australia's Pacific Patrol Boat Replacement Project (SEA 3036) was reached in June 2021. Austal delivered the eleventh 'Guardian' class patrol boat to the Australian Ministry of Defence for onward transfer to the Republic of Kiribati. Named RKS *Teanoai II*, she is one of twenty-one patrol boats being donated to thirteen small Pacific nations as replacements for 'Pacific' class boats supplied under a similar programme in the 1980s and 1990s.[3] Costing some AS$335m (c. US$250m), the project will help Australia's neighbours protect the integrity of maritime zones that extend across vast swathes of the Pacific for less than a tenth of the cost of South Korea's proposed carrier.

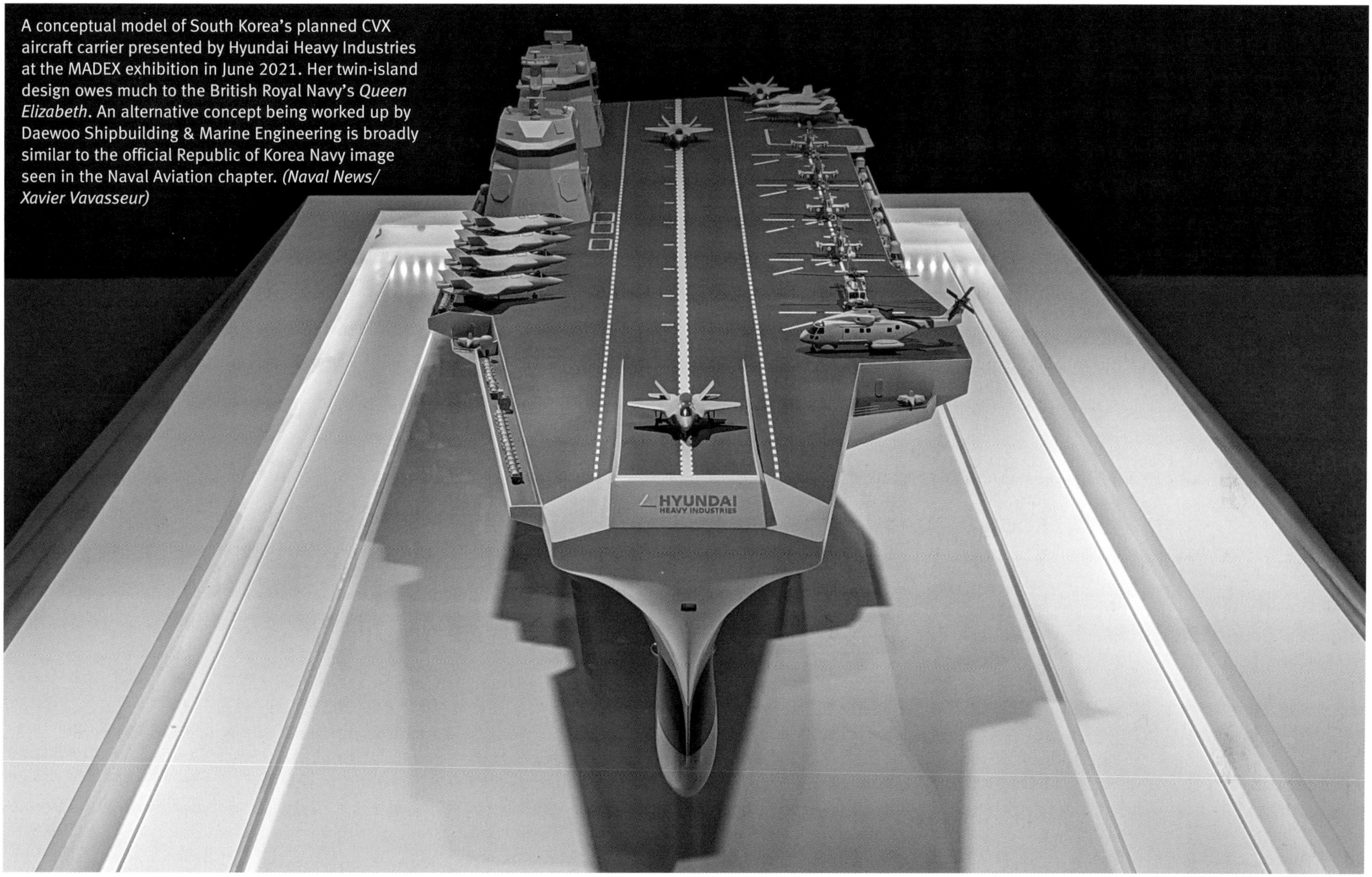

A conceptual model of South Korea's planned CVX aircraft carrier presented by Hyundai Heavy Industries at the MADEX exhibition in June 2021. Her twin-island design owes much to the British Royal Navy's *Queen Elizabeth*. An alternative concept being worked up by Daewoo Shipbuilding & Marine Engineering is broadly similar to the official Republic of Korea Navy image seen in the Naval Aviation chapter. *(Naval News/ Xavier Vavasseur)*

Table 2.2.1: FLEET STRENGTHS IN ASIA AND THE PACIFIC – LARGER NAVIES (MID 2021)

COUNTRY	AUSTRALIA	CHINA	INDONESIA	JAPAN	S KOREA	SINGAPORE	TAIWAN	THAILAND
Aircraft Carrier (CV)	–	2	–	–	–	–	–	–
Support/Helicopter Carrier (CVS/CVH)	–	–	–	4	–	–	–	1
Strategic Missile Submarine (SSBN)	–	7	–	–	–	–	–	–
Attack Submarine (SSN)	–	9	–	–	–	–	–	–
Patrol Submarine (SSK/SS)	6	50	4	21	18	4	4	–
Fleet Escort (DDG/FFG)	11	80	7	36	24	6	26	8
Patrol Escort/Corvette (FFG/FSG/FS)	–	75	24	6	10	14	1	11
Missile Armed Attack Craft (PGG/PTG)	–	75	21	6	18	–	c.30	6
Mine Countermeasures Vessel (MCMV)	4	25	10	20	9	4	c.10	6
Major Amphibious Units (LHD/LPD/LSD)	3	9	6	3	2	4	1	1

Notes: Chinese numbers approximate; some additional Indonesian patrol gunboats are able to ship missiles; Taiwan's submarines are reported to have limited operational availability.

MAJOR REGIONAL POWERS – AUSTRALIA

In July 2020, Australia's government released two major updates to its *2016 Defence White Paper* with publication of its *2020 Defence Strategic Update* and associated *2020 Force Structure Plan*.[4] Whist arguing that the drivers of change identified in the previous strategy were still relevant, the new documents stated that the strategic environment had deteriorated more rapidly than anticipated. The key fear is China's increased use of coercive activities – including 'grey zone' actions just below the threshold of military conflict – to expand its power and influence. The response will be a greater focus of resources on Australia's 'near region' and a sustained commitment to expanding military capabilities by extending planned increased in defence spending throughout the course of the next decade. One eye-catching announcement was the intention to explore the potential for creating a sovereign guided weapons production capability as part of an increased emphasis on precision weaponry.

Australian defence funding is already seeing a healthy level of growth as the government responds to its heightened fears over China's expansion. The military's budget for 2021–2 has been set at AU$44.6bn (c. US$34bn). It is projected to reach AUS$53.3bn (c. US$44bn) by 2024–5. Capability acquisition costs – set to increase from AU$12.7bn in 2020–1 to over AU$20bn by 2024–5 – are driving the uplift. Much of this additional money will be spent on fleet expansion, including the key programmes for new submarines, frigates and offshore patrol vessels that are already underway and which have been little changed as a result of the strategic updates. Having amounted to around AU$1bn p.a. at the end of the last decade, naval shipbuilding costs are expected to ramp up to AU$2.5bn in the current year and are likely to exceed AU$4bn p.a. as future projects move into production.[5]

The then NUSHIP *Supply* approaches her designated base port in Sydney Harbour for the first time in January 2021. She had arrived at Fleet Base West, Western Australia for final fit-out and testing the previous October and was subsequently commissioned on 10 April 2021. *(Australian Department of Defence)*

Table 2.2.2 shows that there have been no significant changes to the Royal Australian Navy's (RAN's) front-line combatant strength since completion of *Hobart* destroyer class deliveries last year. However, capability has been increased by the delivery of the new Navantia-built replenishment oilers *Supply* and *Stalwart* ordered under Project SEA 1654 Phase 3. *Supply* arrived in Australia in October 2020 and was formally commissioned on 10 April 2021. The ceremony attracted national attention for all the wrong reasons due to the invitation of a group of scantily-clad dancers to perform a dance routine that led an anonymous government minister to assert, 'A question worth pondering: what would Horatio Nelson think of this shitshow?'[6] *Stalwart* subsequently reached Australia after her month-long delivery voyage in June 2021 and will be commissioned after the installation of sensitive Australian mission equipment. The arrival of the

Table 2.2.2: ROYAL AUSTRALIAN NAVY: PRINCIPAL UNITS AS AT MID 2021

TYPE	CLASS	NUMBER	TONNAGE	DIMENSIONS	PROPULSION	CREW	DATE
Principal Surface Escorts							
Frigate – FFG	**HOBART** (F-100)	3	6,300 tons	147m x 19m x 5m	CODOG, 28 knots	200	2017
Frigate – FFG	**ANZAC**	8	3,600 tons	118m x 15m x 4m	CODOG, 28 knots	175	1996
Submarines							
Submarine – SSK	**COLLINS**	6	3,400 tons	78m x 8m x 7m	Diesel-electric, 20 knots	45	1996
Major Amphibious Units							
Amph Assault Ship – LHD	**CANBERRA** (JUAN CARLOS I)	2	27,100 tons	231m x 32m x 7m	IEP, 21 knots	290	2014
Landing Ship Dock – LSD	**CHOULES** (LARGS BAY)	1	16,200 tons	176m x 26m x 6m	Diesel-electric, 18 knots	160	2006

two ships should allow the existing tanker *Sirius* to be decommissioned.

The most advanced of the RAN's new programmes is for twelve new *Arafura* class offshore patrol vessels (OPVs) being built to Lürssen's OPV-80 design under SEA 1180 Phase 1. The first two vessels are being completed at the Osborne yard in South Australia but construction has now transitioned to Henderson in Western Australia, where the remainder of the class will be completed by a joint venture between Lürssen and local Australian group Cimvec. A keel-laying ceremony for the third vessel, *Pilbara*, was held at Henderson on 11 September 2020 whilst first steel cutting for the fourth, *Gippsland*, took place the following January. It has also been decided to explore a variant of the design for eight new mine countermeasures 'mother ships' and hydrographic survey vessels set to be acquired under Project SEA 1905 Phase 1, holding out the prospect of an extended production run. Meanwhile, the expected start of *Arafura* class deliveries around the end of 2021 has resulted in the commencement of withdrawal of the current *Armidale* class patrol vessels they will replace. The process began with the decommissioning of *Pirie* at Darwin on 26 March 2021.

The *Armidale* class patrol vessel *Pirie* pictured entering Darwin Harbour for the final time prior to her decommissioning on 26 March 2021. The class is to be replaced by the larger and more capable *Arafura* OPVs over the coming years. *(Australian Department of Defence)*

Although OPV construction is moving from Osborne, its huge shipbuilding complex is set to be the focal point for new frigate and submarine assembly. The *Hunter* class frigate programme – based on the British Type 26 Global Combat Ship design – is the more advanced of the two projects. BAE Systems commenced prototype block fabrication at their ASC Shipbuilding facility at Osborne on schedule in December 2020. The aim is to assemble five representative ship blocks to test production processes and hone workforce skills prior to the start of production of the first actual ship in December 2022. Delivery is now expected in 2029, with eight further units following throughout the 2030s. The cost of this programme is now estimated to amount to AU$45.6bn (c. US$ 35bn) out to 2040, a material uplift on previously-announced costings. Whilst key milestones are being met to date, the size, complexity, risk profile and negative media interest in the contract has resulted in it being added to the Department of Defence's 'projects of interest' watch list.

The *Attack* class diesel-electric submarine project for twelve 'Shortfin Barracuda' variants of Naval Group's *Suffren* class nuclear-powered design appears to be in more trouble. The acquisition under Project SEA 1000 has suffered a storm of criticism relating to local content, delivery schedules and the eventual cost of a programme that is currently budgeted at nearly AU$90bn (c. US$68bn). An indication of the disquiet was provided in an interim report into Australia's Sovereign Naval Shipbuilding Capability by the Australian Senate's opposition-dominated Economics References Committee published in May 2021. Under the chapter heading of 'Submarine Program: Up the Creek without a Canoe' it complained that there had '... been nothing but delays, cost blow-outs, changes of personnel driven by dissatisfaction with the program's outcomes, secret agreements on local industry content way below the level initially promised, and a strained relationship between Defence and Naval Group' since the contract was concluded.[7] It is arguable that this conclusion has been influenced by political partisanship. Indeed, some progress has been made with respect to one major complaint after signature of an agreement in March 2021 that mandates at least 60 percent of the contract's value will be spent in Australia, However, wrangling continues over the expenditure required to design what is, to all intents and purposes, a new submarine. Even if remaining differences are overcome, physical construction of the first new submarine will only start in 2024. All the existing six *Collins* class boats will therefore undergo major mid-life extensions to avoid the emergence of a major capability gap.

MAJOR REGIONAL POWERS – CHINA

The People's Liberation Army Navy's (PLAN's) development over the past twelve months has lacked some of the headline drama of the previous year. In particular, there has been no high-profile commissioning to match that of the first indigenously-built aircraft carrier *Shandong* in December 2019. The underlying picture has, however, largely followed along the same lines as recent trends. High rates of warship construction – including of new, sophisticated types – continue to narrow the qualitative gap with the United States and its allies. At the same time, there has been a continued push to expand China's maritime sphere of influence in the South

China Sea and elsewhere through what Australia's defence update categorised as 'grey zone' actions. An ominous development was the adoption of a new Maritime Police Law in January 2021 that authorised the China Coast Guard to use weaponry against foreign vessels intruding into areas it considers to be under its jurisdiction.[8]

Table 2.2.3 summarises current PLAN fleet strength whilst commentary relating to specific warship types follows below.

Aircraft Carriers and Amphibious Vessels: The assembly of China's first Type 003 aircraft carrier has continued at the Jiangnan shipyard on Changxing Island off Shanghai. Recent satellite images suggest that many of its constituent blocks are now in the course of being integrated. The photographs confirm that the aircraft carrier will be larger than China's two existing vessels of the type and reinforce previous assessments that it will have a conventional CATOBAR (catapult assisted take-off but arrested recovery) configuration. It seems increasingly likely that her launch will take place within the next twelve months, although previous American assessments that the ship will be operational by 2024 may prove to be overly optimistic. Most analysts believe that at least two Type 003 carriers will be constructed and there have been reports that the keel of the second will be laid in the course of 2021.

Meanwhile, China has inducted its first through-deck amphibious assault ship. The lead Type 075 'Yushen' class *Hainan* was commissioned alongside two other major vessels at a ceremony timed to coincide with the PLAN's 72nd anniversary on 23 April 2021. With a reported displacement of over 40,000 tons and a length of 237m, the new ship is only slightly smaller than the US Navy's equivalent *America* (LHA-6) class. Two additional ships are being completed at the Hudong–Zhonghua Shipbuilding yard in Shanghai that built the lead ship. The yard has also finished the latest batch of Type 071 'Yuzhao' class amphibious transport docks with deliveries of two further vessels during the last year. This takes total production to eight units.

Major Surface Combatants: The PLAN has continued the production of multiple types of surface combatant over the past year. At the top end

Table 2.2.3: PEOPLE'S LIBERATION ARMY NAVY: PRINCIPAL UNITS AS AT MID 2021

TYPE	CLASS	NUMBER	TONNAGE	DIMENSIONS	PROPULSION	CREW	DATE
Aircraft Carriers							
Aircraft Carrier – CV	Type 001A (Modified Kuznetsov)	1	65,000 tons	315m x 35/75m x 10m	Steam, 32 knots	Unknown	2019
Aircraft Carrier – CV	Type 001 **LIAONING** (Kuznetsov)	1	60,000 tons	306m x 35/73m x 10m	Steam, 32 knots	Unknown	2012
Principal Surface Escorts							
Destroyer – DDG	Type 055 **NANCHANG** ('Renai')	3	c.12,000 tons	180m x 20m x 7m	COGAG, 30 knots	c.300	2019
Destroyer – DDG	Type 052D **KUNMING** ('Luyang III')	19	7,500 tons	156m x 17m x 6m	CODOG, 28 knots	280	2014
Destroyer – DDG	Type 051C **SHENYANG** ('Luzhou')	2	7,100 tons	155m x 17m x 6m	Steam, 29 knots	250	2006
Destroyer – DDG	Type 052C **LANZHOU** ('Luyang II')	6	7,000 tons	154m x 17m x 6m	CODOG, 28 knots	280	2005
Destroyer – DDG	Type 052B **GUANGZHOU** ('Luyang I')	2	6,500 tons	154m x 17m x 6m	CODOG, 29 knots	280	2004
Destroyer – DDG	Project 956E/EM **HANGZHOU** (Sovremenny)	4	8,000 tons	156m x 17m x 6m	Steam, 32 knots	300	1999
Destroyer – DDG	Type 051B **SHENZHEN** ('Luhai')	1	6,000 tons	154m x 16m x 6m	Steam, 31 knots	250	1998
Destroyer – DDG	Type 052 **HARBIN** ('Luhu')	2	4,800 tons	143m x 15m x 5m	CODOG, 31 knots	260	1994
Frigate – FFG	Type 054A **XUZHOU** ('Jiangkai II')	30	4,100 tons	132m x 15m x 5m	CODAD, 28 knots	190	2008
Frigate – FFG	Type 054 **MA'ANSHAN** ('Jiangkai I')	2	4,000 tons	132m x 15m x 5m	CODAD, 28 knots	190	2005
Frigate – FFG	Type 053 H3 **LIANYUNGANG** ('Jiangwei II')	8	2,500 tons	112m x 12m x 5m	CODAD, 27 knots	170	1992
Frigate – FSG	Type 056/056A **BENGBU** ('Jiangdao')	72	1,500 tons	89m x 12m x 4m	CODAD, 28 knots	60	2013
Plus c. 5 remaining obsolescent frigates of the Type 053 H1/H1G **TAIZHOU/BEIHAI** ('Jianghu II & V') classes that are being steadily retired.							
Submarines							
Submarine – SSBN	Type 094/094A ('Jin')	c. 6	9,000 tons	133m x 11m x 8m	Nuclear, 20+ knots	Unknown	2008
Submarine – SSBN	Type 092 ('Xia')	1	6,500 tons	120m x 10m x 8m	Nuclear, 22 knots	140	1987
Submarine – SSN	Type 093/093A ('Shang')	c. 6	6,000 tons	107m x 11m x 8m	Nuclear, 30 knots	100	2006
Submarine – SSN	Type 091 ('Han')	3	5,500 tons	106m x 10m x 7m	Nuclear, 25 knots	75	1974
Submarine – SSK	Type 039A/039B (Type 041 'Yuan')	17+	2,500 tons	75m x 8m x 5m	AIP, 20+ knots	Unknown	2006
Submarine – SSK	Type 039/039G ('Song')	13	2,300 tons	75m x 8m x 5m	Diesel-electric, 22 knots	60	1999
Submarine – SSK	Project 877 EKM/636 ('Kilo')	12	3,000 tons	73m x 10m x 7m	Diesel-electric, 20 knots	55	1995
Plus c. 10 obsolescent patrol submarines of the Type 035 ('Ming' Class) A Type 032 'Qing' trials submarine has also been commissioned for strategic missile trials.							
Major Amphibious Units							
Amph. Assault Ship – LHD	Type 075 HAINAN ('Yushen')	1	40,000 tons	237m x 36m x 8m	CODAD, 20+ knots	Unknown	2021
Landing Platform Dock – LPD	Type 071 KULUN SHAN ('Yuzhao')	8	18,000 tons	210m x 27m x 7m	CODAD, 20 knots	Unknown	2007

of the displacement scale, the first half of 2021 has seen the commissioning of the second and third c. 12,000-ton Type 055 'Renai' class destroyers. The Jiangnan-built *Lhasa* joined the Northern Fleet in March 2021. She was subsequently followed by the first Dalian Shipbuilding Type 055, *Dalian* – destined for the Southern Fleet – at the same ceremony that saw the commissioning of *Hainan*. A further five vessels are in the course of trials or fitting out at the two shipyards. Both facilities also continue to deliver the smaller, stretched Type 052D 'Luyang III' class design. Around half the twelve units of this variant are now in service in addition to the thirteen original Type 052Ds. Interestingly, however, it has been suggested that PLAN destroyer construction may take a pause after the remaining Type 052Ds and Type 55s have been delivered. The most recent launches of both classes took place at Dalian in August 2020 and satellite imagery has not revealed any signs of further construction.[9] This could suggest that the PLAN wishes to assess the operational performance of its new ships before incorporating lessons learned in a new design.

The much smaller Type 056 'Jingdao' class corvettes also appear to have reached the end of their production run after delivery of *Nanyang* and *Shangqui* in, respectively, January and February 2021. A total of no fewer than seventy-two of these diminutive, 1,500-ton vessels have been completed in two iterations over the eight years from 2013. The first twenty-two, the original Type 056 variant, were followed by fifty of the modified Type 056As, which have the capacity to deploy a towed array. An additional four C13B variants – without anti-submarine capabilities – have been supplied to Bangladesh and two, more basic P18N patrol vessels to Nigeria. The Type 056 hull has also been used as the basis for a more lightly-armed ship used by the China Coast Guard.

Construction of Type 054A 'Jiangkai II' class frigates for the PLAN ended with delivery of the 30th ship early in 2019. However, four of this intermediate type are being completed for Pakistan's navy and there have been several rumours an additional PLAN batch is being built. Some of this speculation refers to a more heavily-armed Type 054X variant with additional VLS cells and, possibly, diesel-electric propulsion. In the meantime, delivery of new vessels is allowing the retirement of the fleet's remaining obsolescent Cold War types. The last two surviving Type 051 'Luda' class destroyers – *Zhanjiang* and *Zuhai* – were decommissioned at a ceremony on 28 August 2020 whilst progressive withdrawals of similar vintage Type 053HI 'Jiangu II' frigates mean that very few of the series remain in operation.

In with the new ... the lead Type 55 destroyer *Nanchang*. Three of these c. 12,000-ton vessels are now in service and a further five under construction, narrowing the PLAN's qualitative gap with the US Navy. *(China Military Online)*

... and out with the old. The Type 051G destroyer *Zuhai* – delivered in 1991 – was decommissioned alongside her sister *Zhanjiang* at a ceremony on 28 August 2020. A comparison with the photograph of *Nanchang* reveals the transformation in PLAN warship design over the past 30 years. *(China Military Online)*

Submarines: The development of the PLAN's secretive submarine arm remains a particularly difficult area to assess. A brief glimmer of light was shed on the Type 094 'Jin' class strategic submarine programme when *Chang-zheng 18* was commissioned alongside *Hainan* and *Dalian* at Yulin Naval Base on Hainan Island in August 2021. She is believed to be one of two additional members of the class completed during the last few years, taking the total of the type in service to six. Published US Naval Intelligence assessments also suggest that a similar number of Type 093/093A 'Shang' nuclear-powered attack submarines are now operational. Production of both designs is likely to be superseded by the next generation Type 95/96 series boats. These are now reportedly being assembled at China's centre of nuclear submarine production at Bohai.[10]

Firm news on the progress of PLAN patrol submarine construction remains sparse, with little known of additional commissionings after a period of rapid deliveries earlier in the decade. However, it seems likely that production of Type 039A/B series of air-independent propulsion (AIP)-equipped 'Yuan' class boats for PLAN service has continued at a low rate alongside export contracts at Wuchang Shipbuilding's facilities given the need to replace the steadily dwindling number of Type 053 'Ming' class submarines. The United States Department of Defense assessed that seventeen 'Yuans' were in service as at the end of 2019. Additionally, May 2021 brought news of a further 'Yuan' class variant with the emergence of photographs showing a boat with a revised fin (sail) similar to that designed for Sweden's A26 *Blekinge* class.

One reason for the apparent hiatus in the expansion of the PLAN's submarine arm may be the wholesale rebuilding and modernisation of the yards responsible for the boats' construction. For example, substantial investment has been in an entirely new facility for patrol submarine assembly at Shuangliu, around 50km downstream of the older shipyard at Wuhan. This may have impacted the rate of production whilst a significant part of the existing yard's production was transferred. A similar process has occurred with respect to production of nuclear-powered submarines, with vast new construction halls at Bohai opened in 2016 as part of a massive revamping of facilities that has not yet been fully completed. These developments have echoed the earlier wholesale transfer of Shanghai's Jiangnan shipyard to Changxing Island in 2009, a move that is now being followed by the shift of the Hudong–Zhonghua facility to the same location.

MAJOR REGIONAL POWERS – JAPAN

The JMSDF's principal units are set out in Table 2.2.4. The Japanese Ministry of Defence's current planning envisages a navy built around a core of twenty-two submarines and fifty-four principal surface combatants, the latter figure including four carrier-like 'helicopter-carrying' destroyers. Submarine numbers are now close to the desired strength, with the commissioning of *Toryu* (SS-512) on 24 March 2021 taking the underwater flotilla to twenty-one boats (plus two training submarines). The twelfth and final member of the *Soryu* class, she will be followed by the slightly larger *Taigei* class. These reportedly contain improved electronics and are the first Japanese submarines to incorporate accommodation for female submariners from build. The lead boat was launched by Mitsubishi Heavy Industries (MHI) at Kobe in October 2020 prior to delivery in March 2022. Three more boats are being built and a fifth was approved in the FY2021 budget. It is interesting to note that the need for twenty-two submarines was first established in 2010, meaning it has taken be over a decade for this to be achieved.

The difficulty in expanding the JMSDF against a backdrop of an essentially flat defence budget is reflected in the surface fleet. This actually saw a slight numerical decline to forty-six units (plus three former surface combatants used as training vessels) year-on-year. The withdrawal of the last two front-line members of the *Hatsuyuki* class and the re-assignment of the final front-line *Hatakaze* class air-defence destroyer *Shimakaze* (DDG-172) to the training role was only partly counterbalanced by the arrival of the new Aegis-equipped *Haguro* (DDG-180). Her arrival takes the number of Aegis destroyers in the fleet to eight, a number which looks set to be further increased following the decision to abandon plans to build two 'Aegis Ashore' ballistic missile defence systems. The government has determined that two Aegis-equipped ships will be acquired in their place. However, it is not known whether or not these are in addition to the current 54-ship target.

The real hope for expanding the fleet rests with the new 30FMM *Mogami* class frigates. Displacing around 5,500 tons at full load, these are smaller than recent JMSDF escort designs. Importantly for a fleet struggling with manpower issues, they can be operated with a crew of as few as ninety personnel. They will, however, remain powerfully armed in spite of their relatively diminutive size and utilise innovations such as a panoramic 360-degree screen incor-

The new JMSDF submarine *Taigei* (SS-513), the lead ship of her class, pictured shortly prior to her launch at Mitsubishi Heavy Industries' shipyard at Kobe on 14 October 2020. *(JMSDF)*

The JMSDF's *Mogami* (FFM-1) class frigate *Kumano* (FFM-2) enters the water at Mitsui Engineering & Shipbuilding's yard at Tamano on 19 November 2020. Delays in *Mogami*'s construction meant that *Kumano* was the first member of the class to be launched. *(JMSDF)*

porating augmented reality know-how in their combat management centre. *Kumano* (FFM-2) was the first ship of the class to enter the water, being launched from Mitsui Engineering & Shipbuilding's yard at Tamano on 19 November 2020. She was followed by the Mitsubishi Heavy Industries Nagasaki-built *Mogami* (FFM-1), which was delayed by an issue with her propulsion plant, on 3 March 2021. Both these ships are expected in service by April 2022. A third frigate, *Noshiro* (FFM-3), was launched by MHI on 22 June 2021.[11] Moreover, all the remaining five vessels from an initial batch of eight ships have either begun construction or have been authorised.

The JMSDF has inducted two other warships in the past year: the third *Awaji* class mine countermeasures vessel *Etajima* (MSO-306) and the third *Hibiki* class ocean surveillance ship *Aki* (AOS-5023). Although the modular *Mogami* class frigates are designed to undertake mine countermeasures tasks,

Table 2.2.4: JAPAN MARITIME SELF-DEFENCE FORCE: PRINCIPAL UNITS AS AT MID 2021

TYPE	CLASS	NUMBER	TONNAGE	DIMENSIONS	PROPULSION	CREW	DATE
Support and Helicopter Carriers							
Helicopter Carrier – DDH	**IZUMO** (DDH-183)	2	27,000 tons	248m x 38m x 7m	COGAG, 30 knots	470	2015
Helicopter Carrier – DDH	**HYUGA** (DDH-181)	2	19,000 tons	197m x 33m x 7m	COGAG, 30 knots	340	2009
Principal Surface Escorts							
Destroyer – DDG	**MAYA** (DDG-179)	2	10,500 tons	170m x 21m x 6m	COGLAG, 30 knots	300	2020
Destroyer – DDG	**ATAGO** (DDG-177)	2	10,000 tons	165m x 21m x 6m	COGAG, 30 knots	300	2007
Destroyer – DDG	**KONGOU** (DDG-173)	4	9,500 tons	161m x 21m x 6m	COGAG, 30 knots	300	1993
Destroyer – DDG	**HATAKAZE** (DDG-171)	0 (2)	6,300 tons	150m x 16m x 5m	COGAG, 30 knots	260	1986
Destroyer – DDG	**ASAHI** (DD-119)	2	6,800 tons	151m x 18m x 5m	COGLAG, 30 knots	230	2017
Destroyer – DD	**AKIZUKI** (DD-115)	4	6,800 tons	151m x 18m x 5m	COGAG, 30 Knots	200	2012
Destroyer – DDG	**TAKANAMI** (DD-110)	5	6,300 tons	151m x 17m x 5m	COGAG, 30 knots	175	2003
Destroyer – DDG	**MURASAME** (DD-101)	9	6,200 tons	151m x 17m x 5m	COGAG, 30 knots	165	1996
Destroyer – DDG	**ASAGIRI** (DD-151)	8	4,900 tons	137m x 15m x 5m	COGAG, 30 knots	220	1988
Destroyer – DDG	**HATSUYUKI** (DD-122)	0 (1)	3,800 tons	130m x 14m x 4m	COGOG, 30 knots	200	1982
Frigate – FFG	**ABUKUMA** (DE-229)	6	2,500 tons	109m x 13m x 4m	CODOG, 27 knots	120	1989
Submarines							
Submarine – SSK	**SORYU** (SS-501)	12	4,200 tons	84m x 9m x 8m	AIP, 20 knots+	65	2009
Submarine – SSK	**OYASHIO** (SS-590)	9 (2)	4,000 tons	82m x 9m x 8m	Diesel-electric, 20 knots+	70	1998
Major Amphibious Units							
Landing Platform Dock – LPD	**OSUMI** (LST-4001)	3	14,000 tons	178m x 26m x 6m	Diesel, 22 knots	135	1998

Note: Figures in brackets refer to trials or training ships.

Although the new *Mogami* class ships are intended to take over some of the JMSDF's mine countermeasures burden, Japan continues to build small numbers of bespoke MCMVs, This is *Etajima* (MSO-306) – the third member of the *Awaji* (MSO-304) class – which was delivered in March 2021. *(JMSDF)*

The ocean surveillance ship *Aki* (AOS-5203) is the third member of the JMSDF's *Hibiki* class. A SWATH design, her hull form is optimised for the deployment of a towed array sonar that can track Chinese and Russian submarines in conjunction with similar US Navy vessels. *(JMSDF)*

continued construction of the *Awaji*s reflects the ongoing priority Japan accords to countering mine warfare and a further member of the class was authorised under the FY2020 budget. Meanwhile, the induction of *Aki*, the first surveillance vessel acquired for nearly thirty years, reflects a desire to achieve enhanced monitoring of PLAN submarine movements as tensions in the Pacific increase. A small-waterplane-area twin hull design, *Aki* has a length of 67m but a beam of 30m and is optimised for the deployment of the US Navy's UQQ-2 Surveillance Towed Array Sensor System (SURTASS). She is similar in design and function to the T-AGOS designated surveillance vessels operated by the US Navy's Military Sealift Command, with both fleets feeding data into the American Integrated Undersea Surveillance System.

In addition to warship acquisitions, Japan continues to authorise slow but steady procurement of F-35B Lighting II strike fighters to operate off the helicopter-carrying destroyers *Izumo* and *Kaga*. Modifications of the two ships for fast jet operation will be carried out in two stages, which will coincide with planned five-year refits. Initial changes to *Izumo* – largely the application of thermal deck protection – were carried out under the FY2020 programme. They will be followed by more extensive changes to her sister *Kaga* – extending to the remodelling of her bow – as part of a programme of work approved in FY 2021. The work on the two ships will then be completed when the ships next go into refit in FY2025 and FY2026, completing the JMSDF's return to a carrier-capable fleet.[12]

MAJOR REGIONAL POWERS – SOUTH KOREA

Although the CVX aircraft carrier has dominated news flow relating to the Republic of Korea Navy over the past year, the programme is just one of numerous projects currently underway that are likely to drive a further significant expansion in fleet capabilities.

Current fleet strength is set out in Table 2.2.5. June 28th 2021 saw the commissioning of the amphibious assault ship *Marado*, a somewhat modified variant of the existing *Dokdo*. Modifications include an enhanced radar suite, the provision of surface-to-air missiles and flight deck strengthening that will allow the operation of heavier aircraft – including the MV-22 Osprey – across a greater area. The fleet has also benefitted from continued progress with the FFX Batch II *Daegu* class frigates, thereby allowing further retirements of the elderly *Ulsan* class light frigate and *Po Hang* corvette types. *Gyeongnam*, the second of eight FFX Batch II vessels, was commissioned in January 2021. Three more are currently fitting out after the launch of *Daejeon* by DSME on 3 May 2021. They will be followed by the FFX Batch III type, on which HHI is already undertaking the detailed design before constructing the lead ship. The new batch will incorporate several improvements but continue to use the Rolls-Royce MT-30 gas turbine that forms the core of the combined diesel-electric and gas (CODLAG) propulsion system in the *Daegus*. The class will eventually comprise six frigates. When combined with the six existing FFX Batch I *Incheon* class vessels already in service, this suggests that South Korea should have an operational fleet of twenty powerful frigates, optimised for littoral warfare but capable of extended deployment, by the end of the decade.

The coming twelve months should see the long-awaited commissioning of the first KSS-III Batch 1 submarine, the DSME-built *Dosan Ahn Changho*, which has been undertaking an extensive series of trials since mid-2019. A second member of the class, *Ahn Moo*, was launched by DSME on 10 November 2020 and HHI is in the course of assembling a third. In a significant enhancement to previous Republic of Korea Navy submarines, the new, c. 3,800-ton boats are equipped with a six-cell VLS for land attack cruise missiles. South Korea plans to construct two additional batches of KSS-III boats in due course, each incorporating improvements over the previous batch. There has been local speculation that at least some of the class will be able to fire (conventionally-armed) ballistic missiles. The navy is also continuing to upgrade its existing KSS-1 *Chang*

The Republic of Korea Navy is making good progress with constructing new frigate and submarine classes into service but deliveries of larger surface combatants are some way into the future. Fabrication of the first of the new KDX-III destroyers commenced in February 2021 for delivery towards the middle of the decade. The class will supplement the existing KDX-III Batch 1 type, of which *Seoae Ryu Seong-ryong* – seen here in August 2020 – is the third and final member. *(US Navy)*

Table 2.2.5: REPUBLIC OF KOREA NAVY: PRINCIPAL UNITS AS AT MID 2021

TYPE	CLASS	NUMBER	TONNAGE	DIMENSIONS	PROPULSION	CREW	DATE
Principal Surface Escorts							
Destroyer – DDG	KDX-III **SEJONGDAEWANG-HAM**	3	10,000 tons	166m x 21m x 6m	COGAG, 30 knots	300	2008
Destroyer – DDG	KDX-II **CHUNGMUGONG YI SUN-SHIN**	6	5,500 tons	150m x 17m x 5m	CODOG, 30 knots	200	2003
Destroyer – DDG	KDX-I **GWANGGAETO-DAEWANG**	3	3,900 tons	135m x 14m x 4m	CODOG, 30 knots	170	1998
Frigate – FFG	FFX-2 **DAEGU**	2	3,600 tons	122m x 14m x 4m	CODLOG	140	2017
Frigate – FFG	FFX **INCHEON**	6	3,000 tons	114m x 14m x 4m	CODOG, 30 knots	140	2013
Frigate – FFG	**ULSAN**	4	2,300 tons	102m x 12m x 4m	CODOG, 35 knots	150	1981
Corvette – FSG	**POHANG**	10	1,200 tons	88m x 10m x 3m	CODOG, 32 knots	95	1984
Submarines							
Submarine – SSK	KSS-2 **SON WON-IL** (Type 214)	9	1,800 tons	65m x 6m x 6m	AIP, 20+ knots	30	2007
Submarine – SSK	KSS-1 **CHANG BOGO** (Type 209)	9	1,300 tons	56m x 6m x 6m	Diesel-electric, 22 knots	35	1993
Major Amphibious Units							
Amph Assault Ship – LHD	LPX **DOKDO**	2[1]	18,900 tons	200m x 32m x 7m	Diesel, 22 knots	425	2007

Notes:
1. *Marado*, the second ship of the class, is built to a slightly different design.

Bogo (German Type 209 design) submarines, with a contract for modification of a second tranche of three boats by DSME announced in December 2020. This tends to confirm previous reports suggesting the underwater flotilla will continue to grow in size as the new KSS-III submarines are inducted. Longer-term ambitions to operate nuclear-powered submarines also appear undiminished.

Future deliveries of larger surface combatants are some way into the distance. In February 2021, HII held a ceremony to mark the first steel-cutting for the lead KDX-III Batch 2 destroyer. One of three planned ships, she will represent a major upgrade over the existing Aegis-equipped *Sejongdaewang-Ham* class by being fitted with ballistic missile defence capabilities. The lead ship is expected to be completed late in 2024 and will be followed by her sisters at biennial intervals. Current plans then envisage production progressing to a KDDX – otherwise known as a KDX-IV – ship. This will be somewhere between the KDX-III and smaller KDX-II type in overall size and will utilise an entirely indigenous combat management system, sometimes referred to as the 'South Korean Aegis'. Tender documents for the ship's basic design were released in 2021. It is hoped to award a construction contract for the first of six ships in 2024, with deliveries from 2030 onwards. In the meantime, continued plans to follow up the modernisation of the existing three KDX-I destroyers with upgrades of the six KDX-II units suggests that South Korea also intends to expand its surface fleet. There has, however, been little further news of the proposed heavily-armed arsenal ship that would be loaded with precision land attack missiles, possibly due to priority being given to the new carrier.

Other developments over the past year have seen the delivery of the South Korean Navy's first dedicated training ship, the HHI-built *Hansando*. Displacing some 4,500 tons, the lightly-armed vessel has an important secondary role as hospital ship. Construction is also progressing well on a second batch of three *Yangyang* class minehunters that are slated for delivery by the end of 2022.

The former Dutch *Van Speijk* class frigate *Karel Satsuitubun* (previously *Isaac Sweers*) was one of the ships deployed to hunt for the missing submarine *Nanggala*. These veteran frigates – based on the *Leander* class – are starting to be retired as new construction joins the Indonesian Navy. *(Australian Department of Defence)*

OTHER REGIONAL FLEETS

Indonesia: It has been a decidedly mixed year for the Indonesian Navy. On 21 April 2021, the Type 209/1300 *Cakra* class patrol submarine *Nanggala* – one of two in service with the Indonesian Navy – was reported missing whilst in the course of completing a torpedo firing exercise some 50 nautical miles off the island of Bali. On 24 April debris believed to have come from the submarine were found in the area. The loss of the boat with all fifty-three personnel onboard was confirmed the following day when wreckage was found at a depth of 853m on the seabed. No conclusive reason for the submarine's loss has yet been established. However, the fact she was some 40 years old at the time of her sinking will likely to have added an increased risk of mechanical failure to the other dangers inherent in underwater operation.

Nanggala's loss came only days after Indonesia commissioned her third and final Type 209/1400 *Nagapasa* class submarine, *Alugoro*, temporarily taking the underwater flotilla to five units. The PT PAL assembled boat – the first submarine completed in Indonesia – was delivered at Surabaya on 17 March and subsequently commissioned at a ceremony on Great Natuna Island in the South China Sea on 6 April 2021. The other two members of the class were completed by DSME, who reached an agreement to supply three additional submarines in 2009. However, there have been subsequent reports that Indonesia is thinking of tearing up the deal in favour of a more modern design. Indonesia is variously reported to have set an overall requirement of between ten and twelve submarines by the end of the decade as it becomes increasingly concerned with overlapping maritime claims with China. Plans for military modernisation revealed in the first half of 2021 that suggest as much as US$125bn will be invested in defence procurement through to the mid-2040s certainly suggest the country is taking this threat seriously.

In June 2021 it became apparent where a significant portion of this money is likely to be spent when Fincantieri announced it had reached an agreement to sell the Indonesian Navy six FREMM type frigates alongside two refurbished former Italian Navy *Maestrale* class. The decision will probably be most disappointing to Damen, who have previously supplied six 'Sigma' type corvettes and frigates, the latter in conjunction with PT PAL. The Indonesian shipbuilder will be involved with the Italian contract, with some reports suggesting it will assemble three of the FREMMs. Once completed,

The Indonesian Navy's brand-new PC-40 type patrol boat *Escolar* photographed in May 2021. Launched by the PT Caputra Mitra Sejati (CMS) shipyard in Java's westernmost province of Banten on 27 July 2020 she is the 15th vessel of a class based on the KCR-40 *Clurit* class FAC. Displacing some 250 tons, her main armament is an Oto Melara Marlin weapons station. *(Australian Department of Defence)*

the programme will provide a considerable boost to Indonesia's small force of front-line combatants, more than compensating for the elderly *Van Speijk* (*Leander*) class frigates that are starting to be retired.

Amongst other developments in the last year, November 2020 saw the start of work on the first of two new minehunters ordered from German shipbuilder Abeking & Rasmussen in January 2019. Based on the Germany Navy Type 332 *Frankenthal* class, it is believed they will replace the existing pair of Dutch 'Tripartite' vessels. Meanwhile work also continues at PT PAL on the fifth and sixth large KCR-60 *Sampari* class fast attack craft, which are scheduled for delivery before the end of 2022.

In spite of the planned boost to its combat forces, the *Tentara National Indonesia* continues to be dominated by its numerous second-line flotilla of patrol and amphibious ships that undertake important constabulary and logistical support duties across the vast archipelago. A wide variety of construction projects remains underway at local shipyards to renew and modernise this fleet. Amongst projects making progress over the last year was the growing *Makassar* series of amphibious transport docks. Tracing their origins to the earlier South Korean built *Dr. Soeharso* (the former *Tanjung Dalpele*), five of these flexible and relatively cheap ships have been delivered for Indonesian Navy from 2007 onwards and others serve in foreign fleets. Their numbers will soon be further expanded by *Dr. Wahidin Soedirohusodo*, the first of two new variants specifically built for the hospital ship role.[13] She was launched by PT PAL in January 2021, a few months after the same yard had commenced work on her as-yet unnamed sister. Other ship launches in the course of the last year include the eighth and ninth *Teluk Bintuni* class tank landing ships, which both entered the water at the PT Bandar Abadi yard on 27 February 2021. At least four additional PC-40 patrol vessel variants of the KCR-40 *Clurit* class missile-armed fast attack craft have also been launched since mid-2020. There will be eighteen of the type in service once they have all been delivered out of a total of forty-two that are ultimately planned.

Malaysia: Malaysia is a good example of one of many South East Asian nations that need to balance the advantages of strong trading links with China against tensions over competing maritime claims. Malaysia's typical approach has been one of quiet diplomacy, a stance perhaps best exemplified by its silence over Chinese harassment of the Malaysian-contracted oil exploration vessel *West Capella* during the course of surveying operations in disputed waters in the first half of 2020. However, in June 2021, the Malaysian government issued a rare public rebuke to China after sixteen People's Liberation Army Airforce aircraft intruded into Malaysian airspace 60 nautical miles north of the state of Sarawak. The incident occurred on 31 May and saw Malaysian Hawk 208 jet trainers being 'scrambled' to intercept the intruders.

In one of the many inconsistencies seen in regional politics, the Royal Malaysian Navy has turned to China's shipbuilding industry to support part of its modernisation efforts. Four *Keris* class patrol ship-like littoral mission vessels were ordered from China Shipbuilding & Offshore International Corporation in 2017, with Wuchang Shipbuilding allocated to undertake the work.[14] The second of these vessels, *Sundang*, was commissioned in March 2021 whilst the as-yet unnamed third and fourth ships were launched in October and December 2020 respectively. Although the project appears to have been completed to time and on budget, Malaysia is looking to other suppliers to supply a second batch of vessels as part of a continuation of a programme that sees an eventual need for eighteen ships of the type. This may reflect the evolving international situation referenced above.

The littoral mission vessel programme forms part of a wider '15 to 5' naval transformation strategy that envisages an eventual focus on just five major warship types. The strategy is making only limited headway in the face of budget constraints and industrial problems. The latter have particularly impacted construction of six larger and more complex 'littoral combat ships' that were ordered from Boustead Naval Shipyard to a variant of Naval Group's 'Gowind' design in 2011. The first ship – launched as *Maharaja Lela* in August 2017 – was due to be delivered in 2019 but the whole construction process has essentially ground to a halt as the shipyard has struggled to complete outfitting work. The Malaysian government reportedly considered cancelling the contract and seeking foreign help to complete the work but, in May 2021, it announced that Boustead would be retained to complete the work. As of mid-2021, there is no clear indication of a strategy to resolve the current problems, nor of when – and at what cost – the contract will eventually be completed. The whole sorry saga has parallels with similar problems afflicting the previous MEKO-100 type *Kedah* class 'new generation patrol vessels', which were also completed by the same facility.

New Zealand: The advent of New Zealand's majority Labour government in the November 2020 elections looks set to alter the future procurement programme established by the previous coalition. In June 2021, the new country's defence minister, Peeni Henare, stated that policy objectives would be reset and overall capital spending reduced in line with the new administration's objectives.[15] Two major projects with significant maritime implications – the acquisition of four P-8A Poseidon maritime patrol aircraft and the mid-life modernisation of the country's two frigates – are already under contract and will not be impacted by the coming budgetary squeeze. However, longer-term plans for an ice-strengthened 'Southern Ocean Patrol Vessel' to complement the existing OPVs and for an amphibious transport dock to provide enhanced sealift and humanitarian response capabilities across the Pacific appear vulnerable to cancellation.

The mid-life 'frigate systems upgrade' (FSU) of the frigates *Te Kaha* and *Te Mana* by Lockheed Martin Canada are now drawing to a close. *Te Kaha* returned to New Zealand in December 2020 after completion of the upgrade package in Victoria, British Columbia and is expected to return to operational status by 2022 after completing further work in New Zealand and a series of post-refit trials and exercises. *Te Mana* followed her sister into dockyard hands in May 2019. Although industrial work is now complete, post-refit testing has been extended by the pandemic and her return home may be delayed.

The FSU project – encompassing a new Lockheed Martin CMS 30 combat management system, Thales SMART-S Mk 2 radar and Spherion B sonar, MBDA Sea Ceptor surface-to-air missiles and various other sensor and countermeasures upgrades – proved to be more difficult and costly than initially appreciated. This resulted in previous plans for a new littoral operations support vessel being downgraded to the second-hand acquisition of the former Norwegian offshore support vessel *Edda Fonn* to provide diving support, hydrographic and salvage functions. The ship was commissioned as *Manawanui* in June 2019, taking the name of a previous vessel. However, a series of upgrades and capability enhancements has subsequently been undertaken in New Zealand prior to planned 'operational release' in the second half of 2021 following completion of tests and trials.

The Royal New Zealand Navy's *Anzac* class frigate *Te Kana* completed her midlife systems upgrade in the course of 2021, returning to New Zealand in December 2021. The upgrade of her sister-ship *Te Mana* is now in the post-refit testing and trials phase. *(New Zealand Ministry of Defence)*

North Korea: There has been little tangible news flow relating to the Democratic People's Republic of Korea Navy over the past year in spite of continued speculation on the future direction of the country's strategic submarine programme. In late 2020, reports emerged that North Korea might be constructing a third, larger ballistic missile-carrying submarine. This would be capable of carrying a greater missile load than either the existing sole 'Sinpo B' or 'Gorae' class trials boat or the existing modified Type 033 'Romeo' class boat that was revealed in video footage released in mid-2019. The modified 'Romeo' – sometimes referred to as a 'Sinpo C' – has yet to enter service and there are conflicting assessments as to whether this is designed for the operational deployment of North Korea's Pukguksong-3 (KN-26) submarine-launched ballistic missile or to facilitate trials of the subsequent Pukguksong-4 and Pukguksong-5 strategic missiles that have been displayed at recent military parades. North Korean leader Kim Jong-un has also claimed that the country is in the early stages of developing a nuclear-powered submarine to give greater effect to the country's steadily-expanding nuclear deterrence capability.[16]

The Philippines: The Philippine Navy has continued to make meaningful progress with its modernisation efforts under the so-called 'Horizon 2' (i.e. 2018–22) phase of the country's defence modernisation programme in spite of reductions in procurement spending driven both by the COVID-19 pandemic and competing budget priorities. Notably, the last year has seen the formal commissioning of the two new light frigates acquired from South Korea's HHI. *Jose Rizal* (FF-150) was inducted at a ceremony held in Subic Bay on 10 July 2020 after a delay caused by a need to isolate her crew when one was infected by the pandemic. In December she led a parade of navy ships – also including the modern Indonesian-built amphibious transport docks *Tarlac* (LD-601) and reviewing vessel *Davao del Sur* (LD-602) – in a rare fleet review to commemorate the 85th anniversary of the founding of the Philippine Armed Forces. Sister-ship *Antonio Luna* (FF-151) subsequently arrived in Philippine waters in February 2021 and was commissioned on 19 March of that year.

A number of projects remain outstanding under 'Horizon 2'. In May 2021, it was confirmed that a contract had been signed with Israel Shipyards earlier in the year to acquire a total of nine 'Shaldag Mk V' fast patrol boats to meet a requirement for fast interceptor craft. At least four of the vessels will be armed with Spike non-line-of-sight (NLOS) missiles. The deal also includes provision for the transfer of knowledge to allow the construction of some of the patrol boats and their subsequent maintenance in a Philippine yard. The vessels will replace a flotilla of eight South Korean-built 'Sea Dolphin' type fast attack craft. These have already been retired as part of a 'Phase In, Phase Out' programme that is also seeing the withdrawal of the fleet's final Second World War era vessels. It also seems probable that an agreement is imminent for the construction of six Austal-designed 80m OPVs that would be enlarged from the group's existing 'Cape' class and built at their yard in Balamban, Cebu. However, plans for additional amphibious transport docks – as well as the navy's ambitious submarine acquisition project – now seem likely to be deferred to the 'Horizon 3' (2023–8) stage of the modernisation programme.

Singapore: The end of August 2020 saw ThyssenKrupp Marine Systems (TKMS) commence trials of *Invincible*, the first of four Type 218SG submarines ordered in two batches in 2013 and 2017. Equipped with AIP and broadly similar in size

The Philippine Navy received its first-ever modern frigate with the commissioning of the HHI-built *Jose Rizal*, a derivative of the South Korean Navy's FFX series of designs, on 10 July 2020. Her sister, *Antonio Luna*, was inducted into the fleet on 19 March 2021. *(US Navy)*

to Israel's 'Dolphin 2' class submarines also built by the German company, the new boat is expected to be delivered in the first half of 2022.

December 2020 marked the end of an era with the decommissioning of *Freedom* and *Gallant*, the final two *Fearless* class patrol vessels in front-line service. The first warship class both designed and built in Singapore, their role has been assumed by the new *Independence* class littoral mission vessels. However, four of the class quickly gained a new lease of life as *Sentinel* class maritime security and response vessels, forming the core of a new Maritime Security and Response Flotilla that was stood up on 26 January 2021. Part of a restructured Maritime Security Command, the ships will support constabulary operations across Singapore's littoral waters until purpose-built vessels arrive in 2026.

The new manned warships will be supplemented by new Maritime Security Unmanned Surface Vessels. These have been developed by the Singapore Ministry of Defence in partnership with the country's Defence Science & Technology Agency procurement body and the DSO National Laboratories research organisation. Displacing around 30 tons and operated autonomously under the supervision of two operators, they can be equipped with a range of non-lethal and lethal systems including strobe and search lights, a long range acoustic device and a 12.7mm stabilised weapon system with laser dazzler. Autonomous navigation systems and an indigenously-developed collision detection and avoidance system are intended to ensure that the vessels can navigate through the busy traffic of the Singapore Strait whilst coping with the constantly changing conditions at sea. In March 2021, Singapore's Minister for Defence, Dr Ng Eng Hen stated that the new vessels should complete trials by the end of the year before starting to complement their manned counterparts at sea.

One of the Republic of Singapore Navy's new Maritime Security Unmanned Surface Vessels. Capable of being operated autonomously under the supervision of two shore-based operators, it is expected that the new vessels will be deployed operationally before the end of 2021. *(Republic of Singapore Navy)*

Taiwan: Taiwan's Republic of China Navy is starting to make material progress with procurement efforts after a long period of quantitative and qualitative decline against China's PLAN. Whilst by no means restoring the gap between the two navies, the delivery of vessels now under construction could seriously complicate any Chinese effort to unite the island with the rest of the mainland by force.

The most significant recent event has been the commencement of work on the first of eight indigenous patrol submarines on 24 November 2020. This was marked by a high-profile ceremony at the purpose-built CSBC Corporation facility in the southern port city of Kaohsiung that will build the new boats. Scheduled to enter service in 2025, the c. 2,500-ton submarine is said to be based on the navy's existing pair of Dutch-built modified *Zwaardvis* class and incorporate significant amounts of American and European equipment. In March 2021, the United States' Biden administration approved the export of vital components that are reported to include sonar, combat management and periscope systems.[17] Taiwan will also need to source torpedo discharge equipment, technology that is currently produced by only a handful of nations.

Other developments over the past year have included the launches of the first improved *Tuo Chinag* catamaran-hulled corvette on 15 December 2020 and of *Yu Shan*, the lead ship of a planned quartet of amphibious transport docks on 13 April 2021. The new corvette, named *Ta Chiang*, subsequently commenced sea trials early in May. Reportedly capable of speeds of up to 45 knots, a total of twelve of these stealthy vessels should be delivered by the mid-2020s as Taiwan bolsters its asymmetric naval capabilities. The delivery of the new fast minelaying ship *Min Jiang* – first of a class of four – in December 2020 is further evidence of the country's evolving anti-access/area denial (A2/AD) strategy.

Thailand: The Thai defence budget is under some pressure as a result of the current pandemic. One victim has been the Royal Thai Navy's controversial programme to acquire three Chinese-manufactured S26T submarines based on the PLAN's 'Yuan' class. The first unit was ordered in 2017 but plans for two more boats have been shifted backwards, partly in response to public opposition. The submarine currently under construction is expected to be delivered by 2023. By that time, Thailand should also have received its Type 071E 'Yuzhao' class amphibious transport dock, the other major naval procurement currently underway.

Indigenous shipbuilding continues to be dominated by the construction of mid-sized coastal patrol boats of a number of different designs varying between 36m and 42m in length by the Marsun group. Two of the smaller type – *Tor 114* and *Tor 115* – were delivered in March 2021 whilst two slightly larger ships – *Tor 997* and *Tor 998* – were launched in June. There has also been speculation that local construction of a third 'River Batch 2' OPV by Bangkok Dock will be approved in the next defence budget, ultimately following *Krabi* and *Prachuap Khiri Khan* into service.

Vietnam: A few years have now passed since the Vietnam's People's Navy completed a major acquisition programme that ultimately encompassed six Russian submarines and four frigates. Although there has been no firm news, there is speculation that a further round of investment is imminent. This could include additional, improved Project 11661E 'Gepard' type frigates and modular Project 22160 *Bykov* class large patrol ships. In the interim, the security of the new submarine flotilla and its personnel has been boosted by the delivery of the submarine rescue ship *Yet Kieu*, which was built by the local Z189 Shipyard to a Damen design previously supplied to support Royal Australian Navy operations. The new ship is reported to use a package of British-supplied remotely-operated vehicles and rescue equipment. Further diversification of supply is evidenced by the arrival of the former US Coast Guard high endurance cutter *John Midgett* (WHEC-726), which will join her sister *Morgenthau* (WHEC-722) in Vietnam Coast Guard service. In July 2020, an agreement was reached with Japan for the supply of six large patrol vessels, further reinforcing Vietnam's ability to police disputed waters in the South China Sea.

Notes

1. Up until 2018, the Republic of Korea's deterrence strategy was described as being based on a triad of capabilities, viz. (i) a 'Kill Chain' pre-emptive strike system aimed at attacking relevant North Korean installations before missiles are launched, (ii) a Korea Air & Missile Defence (KAMD) to destroy incoming missiles and (iii) a Korea Massive Punishment & Retaliation targeting the North Korean leadership in Pyongyang. In that year, South Korea's Moon Jae-in administration combined and renamed elements (i) and (iii) as the Strategic Strike System, possibly reflecting a desire to reduce tensions on the peninsula. A comprehensive description of the strategy is provided by Ian Bowers and Henrik Stålhane Hiim, 'Conventional Counterforce Dilemmas: South Korea's Deterrence Strategy and Stability on the Korean Peninsula' in *International Security* – Volume 45, Issue 3 (Cambridge MA: MIT Press, 2021), pp.7–39.

2. An excellent review of the South Korean Navy's expansion into one of the world's leading fleets is provided by Ian Bowers, *The Modernisation of the Republic of Korea Navy: Seapower, Strategy and Politics* (London: Palgrave Macmillan, 2019).

3. The countries benefiting from the Pacific Patrol Boat Replacement Project are the Cook Islands, the Federated States of Micronesia, Fiji, Kiribati, the Marshall Islands, Palau, Papua New Guinea, Samoa, the Solomon Islands, Timor Leste, Tonga, Tuvalu and Vanuatu. The 39.5m steel-hulled boats are faster and have better sea-keeping features than the boats they replace. All should be delivered before the end of 2023.

4. *2020 Defence Strategic Update* and *2020 Force Structure Plan*, both (Canberra: Australian Department of Defence, 2020) and available at: www1.defence.gov.au/strategy-policy/strategic-update-2020

5. A comprehensive review of Australian defence spending is contained in *The Cost of Defence: ASPI Defence Budget Brief 2021-2022* (Canberra: Australian Strategic Policy Institute, 2021) and available on the aspi.org.au website.

6. The unnamed minister was quoted in an article by Andrew Greene, 'Military reminded core business is to use "lethal violence" to defend Australia's values and sovereignty' posted to the *ABC News* website – www.abc.net.au/news – on 13 April 2021.

7. See The Senate Economics Reference Committee, *Australia's Sovereign Naval Shipbuilding Capability: Interim Report* (Canberra: Commonwealth of Australia, 2021), p.22. An interesting aspect of the committee's deliberations was its observations on competing bids for the programmes, not least German TKMS' offer of a AU$20bn fixed price contract to build Australia's submarines. As noted by the committee, '*Das Boot*, however, has sailed'.

8. See Gabriel Dominguez, 'China authorises coast guard's use of firepower against foreign vessels', *Jane's Defence Weekly* – 5 February 2021 (Coulsdon: Jane's, 2021), p.4.

9. This assessment was provided by Sean O'Connor in 'PLAN DDG production hits pause', *Jane's Defence Weekly* – 17 March 2021 (Coulsdon: Jane's, 2021), p.30.

10. The assessment of current PLAN submarine strength is taken from the US Department of Defense's *Military and Security Developments Involving the People's Republic of China 2020: Report to Congress* (Washington DC: Office of the Secretary of Defense, 2020), p.45. An update on future PLAN nuclear-powered submarine construction was provided by I H Sutton in, 'First Image Of China's New Nuclear Submarine Under Construction' Posted to the *Naval News* website – navalnews.com – on 1 February 2021.

11. In March 2021 it was announced that MHI would acquire Mitsui Engineering & Shipbuilding's naval business, meaning that all the *Mogami* class will be built by the same group. Three Japanese companies now remain active in naval shipbuilding, the others being Japan Marine United and Kawasaki Heavy Industries.

12. The work packages are outlined in the *Defense Programs and Budget of Japan* FY2020 and FY2021 budget overviews, both available at: mod.go.jp. For further detail see Kosuke Takahashi, 'Japan's *Izumo*-class carriers will not feature a "ski-jump" ramp for F-35B operations', *Jane's Defence Weekly* – 21 October 2020 (Coulsdon: Jane's, 2021), p.6.

13. *Dr. Soeharso* was reconfigured as a hospital ship and the fifth *Makassar*, *Semarang*, is temporarily operating in this role. However, the two ships currently under construction were seemingly ordered with this particular mission in mind.

14. Two of the vessels were originally scheduled to be built in Malaysia at the Boustead Naval Shipyard. However, possibly because of the yard's problems completing littoral mission ships, construction of all four ships ultimately took place in China.

15. See 'Covid-19 forced govt to cut back defence spending - Peeni Henare' posted to the RNZ (Radio New Zealand) website – rnz.co.nz – on 25 June 2021.

16. Two sites continue to provide useful regular updates on North Korea's strategic submarine programme. These are The Henry L. Stimson Center think tank's *38 North* project at www.38north.org and defence analyst H I Sutton's *Covert Shores* site at www.hisutton.com

17. Details of the American equipment being supplied were listed in Lawrence Chung's 'Biden gives Taiwan a boost with submarine technology it can't build on its own' posted to the *Insider* website – www.businessinsider.com – on 17 March 2021

2.3 REGIONAL REVIEW

Author:
Conrad Waters

THE INDIAN OCEAN AND AFRICA

Whereas the Asia-Pacific region and Europe are increasingly the focal points of great-power tensions, the crucible of actual conflict remains centred on the Middle East and North Africa. Here, recent years have seen hostilities of varying intensity fought out on battlegrounds as far apart as Libya in the west and Yemen in the south. Although these wars have tended to have local origins, they have largely been sustained through the involvement of regional and external powers seeking to extend their influence. Whilst the traditional dividing lines of Arab-Israeli and Shia-Sunni antagonism continue to have relevance, a fluctuating and frequently confusing network of friendships and enmities is seemingly driven more by national interests than broader cultural or religious affiliations.[1] The Turkish Erdoğan presidency's drive to increase its own influence across the Middle East – typified by active intervention in Libya and Syria and seen by some as harking back to the imperialism of the Ottoman Empire – has been just one further addition to an unstable mix.

Although much of the fighting in the region has been land-based in nature – including the 'largest drone war in the world' over the skies of Libya – Middle Eastern hostilities have pertinence to naval developments on a number of levels.[2] Most notably, fears over both Iranian and Turkish expansion are driving a considerable surge in naval investment, respectively in the Persian Gulf and along the Mediterranean's southern shores. This trend has been most apparent in Egyptian naval procurement. Here, previous reports of major naval acquisitions from Italy have been confirmed with the rapid arrival of two FREMM type frigates as the first instalment of what is widely reported to be a much larger deal. Their arrival marks a further stage in a remarkable programme of modernisation and expansion since the commencement of the Sisi presidency in 2013, largely considered to have been financed with the help of support from allies in the Arab world.[3]

Events in the Middle East also have the potential to have an impact on the broader global naval balance by enabling the Russian Navy to rebuild a presence in the region to an extent not seen since the height of the Cold War. This process commenced following Russia's support of the Assad regime in Syria's civil war, providing the pretext for the expansion of residual Soviet-era naval facilities at the port of Tartus. Alongside similar aviation facilities at the Hmeimim Air Base south of Latakia, these are home for missile batteries considered capable of providing Russia with a powerful anti-access/area-denial (A2/AD) 'bubble' over the Eastern Mediterranean. In November 2020 it emerged that Russia had reached agreement with Sudan to establish a naval logistical hub at Port Sudan on the Red Sea, further strengthening its footprint close to one of the world's major maritime arteries.[4] Russia has also been a supporter of the Libyan National Army faction in Libya's recent civil war. Here the Wagner Group – a private military company with strong links to the Russian defence establishment – has been particularly active. Some commentators believe that this support was also a precursor to what now seems an abortive attempt to obtain further overseas naval basing rights, perhaps at Tobruk or Benghazi.

Russia's regional expansion comes at a time when events in the Suez Canal have highlighted the vulnerability of trade between Europe and the Far East to potential interdiction. On 23 March 2021, the Taiwanese Evergreen Marine-operated container ship *Ever Given* grounded in the canal some five nautical miles north of Suez. The waterway was completely blocked for the following six days until a substantial salvage effort finally released the stricken vessel. Several hundred ships faced delays transiting the Canal, with as much as US$10bn of goods held up for each day the *Ever Given* remained grounded.

Russian ambitions in the Mediterranean and wider Middle East appear to be more about promoting its influence abroad and counterbalancing the current dominance of NATO forces than actually confronting the 'Western' allies or directly threatening their maritime trade routes.[5] However, it is noteworthy that Russia's presence in the Middle East is on the rise at the same time Washington is looking to pare back its own regional footprint. All this points to some of the hard choices the United States faces – as well as the importance of its attempts to rejuvenate its own network of alliances – as it looks to rebalance its strength in the face of increasing strategic challenges.

Table 2.3.1 provides an overview of the most significant regional navies as of mid-2021.

The Egyptian FFG-7 type frigate *Sharm El Sheikh* – formerly USS *Fahrion* (FFG-22) – and the modified *Mistral* class amphibious assault ship *Anwar El Sadat* pictured whilst exercising with British Royal Navy warships in November 2020. The Egyptian Navy has undergone a remarkable expansion in recent years and continues to acquire new vessels. *(Crown Copyright 2020)*

Table 2.3.1: FLEET STRENGTHS IN THE INDIAN OCEAN, AFRICA AND THE MIDDLE EAST – LARGER NAVIES (MID 2021)

COUNTRY	ALGERIA	EGYPT	INDIA	IRAN	ISRAEL	PAKISTAN	SAUDI ARABIA	SOUTH AFRICA
Aircraft Carrier (CV)	–	–	1	–	–	–	–	–
Strategic Missile Submarine (SSBN)	-	-	1	-	-	-	-	-
Attack Submarine (SSN/SSGN)	–	–	–	–	–	–	–	–
Patrol Submarine (SSK/SS)	6	7	14	4	5	5	–	3
Fleet Escort (DDG/FFG)	2	11	22	-	-	7	7	4
Patrol Escort/Corvette (FFG/FSG/FS)	10	5	12	8	3	2	4	–
Missile Armed Attack Craft (PGG/PTG)	c.12	c.30	8	c.25	8	9	9	2
Mine Countermeasures Vessel (MCMV)	2	c.14	–	1	–	3	3	4
Major Amphibious (LPD)	1	2	1	–	–	–	–	–

Notes:

1 Algerian fast attack craft and Egyptian fast attack craft and mine-countermeasures numbers approximate.

2 Iranian fleet numbers exclude large numbers of indigenously-built midget and coastal submarines, as well as numerous additional missile-armed patrol boats operated both by the Iranian Navy and Revolutionary Guard .

3 The South African attack craft and mine-countermeasures vessels serve in patrol vessel roles.

INDIA

India's defence budget has grown to become the third largest in the world, seeing a real-terms uplift of over a third in the past decade. However, much of this total has been swallowed up by personnel costs – particularly sums due to some 3.2 million pensioners – of India's numerically vast defence establishment. This has meant that there has been little to spare for higher capital expenditure on equipment, which shrank from 27 percent to just 19 percent of total Indian Ministry of Defence spending between 2010–11 and 2020–1. There have been recent attempts to reverse this trend, with projected modernisation spending scheduled to increase by 23 percent in 2021–2 within an only slightly enlarged total defence budget. The Indian Navy's share of the total has been set at INR322bn (c. US$4.3bn), a 26 percent uplift on last year's initial estimate of INR256bn (c. US$3.5bn). The problem is that this figure is insufficient to cover a projected annual capital expenditure requirement for authorised programmes (as at 2020–1) of c. INR409bn (c. US$5.5bn). In practice, this has meant that some additional money has to be found on a regular basis to supplement the original estimates to complete priority acquisitions, with other programmes cancelled or deferred.[6]

The Indian Navy's prospects have also not been helped by the appointment of Indian Army General Bipin Rawat as India's first Chief of Defence Staff on 1 January 2020. Perhaps understandably given his background, the general is viewed as wanting to prioritise the defence of India's land borders with China and Pakistan, leaving some future naval programmes in doubt. There is particular scepticism about the Indian Navy's long-held desire to field three aircraft carriers, which would allow one to be operational on each of the east and west coasts whilst the third is in refit. Instead, it is argued that limited funds should be focused on renewing the steadily-degrading submarine flotilla. In early June 2021, long awaited authorisation was given to proceed with tenders for six Project 75I submarines, suggesting plans for the third carrier may remain in abeyance for the time being unless more money can be found.

The Indian Navy is starting to retire its *Rajput* class destroyers. *Ranvijay* – pictured here in November 2020 – was, however, the last member of the class to be completed and may be good for a few more years. *(US Navy)*

Current Indian Navy composition is provided in Table 2.3.2, with developments relating to major types summarised below:

Aircraft Carriers and Major Amphibious Ships: Completion of the long-awaited *Vikrant*, India's first indigenously-built aircraft carrier, has been further delayed by the COVID-19 pandemic. She reportedly completed basin trials at Cochin Shipyard Ltd (CSL) at the end of November 2020 and is now expected to go to sea for the first time in the summer of 2021. Her subsequent period of trials and working-up is likely to be lengthy, suggesting it may not be until the middle of the decade that the new ship is fully operational.

As mentioned above, the navy's hopes for a third carrier to join *Vikrant* and the existing Russian-built *Vikramaditya* are facing heavyweight opposition. As such, it seems more likely that the planned ship will become a replacement for *Vikramaditya* if the project eventually proceeds. Also in abeyance are plans for four 'amphibious transport docks' that had actually expanded into much larger LHD-type amphibious assault ships based on either the French *Mistral* or Spanish *Juan Carlos I* designs. Reports suggest a longstanding request for proposals was finally withdrawn after financial difficulties at two of the three potential bidders left only a Larsen & Toubro offer based on the Spanish ship in the running. The specification for the new ships will now be revised against a backdrop of financial constraints that may see the programme reduced to just two vessels.

Major Surface Combatants: The last year has seen the Indian Navy's destroyer flotilla temporarily fall to nine vessels with the withdrawal of the lead Project 61ME class vessel *Rajput*. First commissioned in May 1980, the destroyer left the fleet after 31 years of service at a ceremony held on 21 May 2021. Three members of the five-ship class now remain in service. *Ranjit* – decommissioned in 2019 – was sunk as a target in the course of the TROPEX-21 exercise in February 2021.

The Project 61MEs will, in practice, be replaced by the four new 8,100-ton Project 15B *Visakhapatnam* class destroyers which remain under construction at Mazagon Dock Shipbuilders Ltd (MDSL). The first ship is believed to have finished basin trials at the end of 2020 and the shipbuilders remain optimistic she can be delivered by the end of

Table 2.3.2: INDIAN NAVY: PRINCIPAL UNITS AS AT MID 2021

TYPE	CLASS	NUMBER	TONNAGE	DIMENSIONS	PROPULSION	CREW	DATE
Aircraft Carriers							
Aircraft Carrier (CV)	Project 1143.4 **VIKRAMADITYA** (KIEV)	1	45,000 tons	283m x 31/60m x 10m	Steam, 30 knots	1,600	1987
Principal Surface Escorts							
Destroyer – DDG	Project 15A **KOLKATA**	3	7,400 tons	163m x 17m x 7m	COGAG, 30+knots	330	2014
Destroyer – DDG	Project 15 **DELHI**	3	6,700 tons	163m x 17m x 7m	COGAG, 32 knots	350	1997
Destroyer – DDG	Project 61 ME **RAJPUT** ('Kashin')	3	5,000 tons	147m x 16m x 5m	COGAG, 35 knots	320	1980
Frigate – FFG	Project 17 **SHIVALIK**	3	6,200 tons	143m x 17m x 5m	CODOG, 30 knots	265	2010
Frigate – FFG	Project 11356 **TALWAR**	6	4,000 tons	125m x 15m x 5m	COGAG, 30 knots	180	2003
Frigate – FFG	Project 16A **BRAHMAPUTRA**	3	4,000 tons	127m x 15m x 5m	Steam, 30 knots	350	2000
Frigate – FFG	Project 16 **GODAVARI**	1	3,850 tons	127m x 15m x 5m	Steam, 30 knots	315	1983
Corvette – FSG	Project 28 **KAMORTA**	4	3,400 tons	109m x 13m x 4m	Diesel, 25 knots	195	2014
Corvette – FSG	Project 25A **KORA**	4	1,400 tons	91m x 11m x 5m	Diesel, 25 knots	125	1998
Corvette – FSG	Project 25 **KHUKRI**	4	1,400 tons	91m x 11m x 5m	Diesel, 25 knots	110	1989
Submarines							
Submarine – SSBN	**ARIHANT**	1	7,500+ tons	112m x 11m x 10m	Nuclear, 25+ knots	100	2016
Submarine – SSK	Project 75 **KALVARI** ('Scorpène')	3	1,800 tons	68m x 6m x 6m	Diesel-electric, 20 knots	45	2017
Submarine – SSK	Project 877 EKM **SINDHUGHOSH** ('Kilo')	7	3,000 tons	73m x 10m x 7m	Diesel-electric, 17 knots	55	1986
Submarine – SSK	**SHISHUMAR** (Type 209)	4	1,900 tons	64m x 7m x 6m	Diesel-electric, 22 knots	40	1986
Major Amphibious Units							
Landing Platform Dock – LPD	**JALASHWA** (AUSTIN)	1	17,000 tons	173m x 26/30m x 7m	Steam, 21 knots	405	1971

2021. If this target is achieved, she will have been completed in around eight years, some two years faster than the later members of the previous and broadly similar Project 15A *Kolkata* class.

Frigate construction remains dominated by the Project 17A *Nilgiri* class ships split between MDSL (four units) and Kolkata's Garden Reach Shipbuilders & Engineers or GRSE (three units). Six of the frigates were under construction as of mid-2021 with GRSE's first unit – *Himgiri* – joining her MDSL sister *Nilgiri* in the water following her launch on 14 December 2020. Displacing around 6,700 tons, the new frigates will be upgraded, stealthier variants of the previous Project 17 *Shvalik* design incorporating the EL/M-2248 MF-STAR phased array found on the larger destroyers. It remains to be seen whether scheduled delivery dates from 2022 to 2025 are realistic.

Work also continues on the programme for four modified Project 11356 *Talwar* class frigates signed with Russia at the end of 2018. Two of these are ships originally under construction for the Russian Navy but suspended after Ukraine cut off the supply of their gas turbines due to Russia's annexation of the Crimea. India has subsequently concluded an agreement to acquire the turbines directly from Ukraine and these have now been delivered to the Yantar Shipyard in Kaliningrad that is building the two ships. A second pair, reportedly with a modified equipment fit, are being built under licence by Goa Shipyard in India. The facility laid the keel of its first ship on 29 January 2021 and her sister on 18 June 2021. Deliveries are expected to be concluded in the first half of 2027. This will bring the total number of Project 11356 variants in Indian Navy service to ten vessels.

Submarines: Although the agreement to seek tenders for the new Project 75I submarines is a positive development, the largely elderly Indian submarine flotilla remains under pressure. The transfer of the Project 877EKM 'Kilo' or *Sindhughosh* class submarine *Sindhuvir* to Myanmar in 2020 is expected to be followed by the withdrawal of the 1987-built *Sindhudhvaj* within the next few months. Modernisation will allow some of the remaining seven boats to be retained in service for a few more years. Upgrades are also being undertaken or planned on the quartet of Type 209 *Shishumar* class submarines that were first commissioned between 1986 and 1994. However, the only truly modern patrol submarines are the six 'Scorpène' type boats ordered from MDSL under licence from France's Naval Group. The third of these, *Karanj*, was commissioned on 10 March 2021 and the fifth, *Vagir*, launched the previous November. All six should be in service by 2023. It is likely that there will then be a considerable delay in new deliveries until the Project 75I programme comes good, possibly around the end of the decade.

The Indian Navy is widely reported to have an eventual requirement for twenty-four non-strategic submarines, six of which could be nuclear-powered. Clearly, this vision is unlikely to be achieved on the basis of current and planned construction. Indeed, India's operation of nuclear-powered submarines has suffered a setback with the return to Russia of the leased Project 971 'Akula' class boat *Chakra* in June 2021, several months before the agreement was due to end. A quickly withdrawn report by the Russian news agency TASS suggested the lease's premature termination followed the explosion of a high-pressure air cylinder onboard the submarine in the spring of 2020 that damaged the double hull.[7] India had previously agreed the lease of another 'Akula' to take effect from the middle of the decade but the gap could impact the training of crews for new *Arihant* class strategic submarines that are set to form the

The new Indian 'Scorpène' type submarine *Vagir* – fifth member of India's *Kalvari* class – pictured in the course of launch from MDSL in Mumbai on 12 November 2020. Although good progress is now being made in delivering the long-awaited six-strong class, the outlook for India's submarine flotilla is far from encouraging. *(Naval Group)*

cornerstone of India's nuclear deterrent. *Arighat*, the second of these secretive boats, has been variously reported as undergoing harbour acceptance tests or preliminary sea trials as of early 2021 and could be delivered before the year's end.

Other Ships: Often associated with the strategic submarine programme are two new missile-tracking vessels that will be operated by the Indian Navy for the country's National Technical Research and Defence Research & Development Organisations. One, named *Dhruv*, was reportedly delivered to the navy by Hindustan Shipyard in October 2020 whilst another smaller and as yet unnamed vessel remains under construction at the CSL. In addition to supporting Indian missile trials, the vessels are also expected to play a part in ongoing ballistic missile defence efforts.

Turning to green water operations, March 2021 saw the commissioning of *L58*, the eighth and final member of a class of 1,000-ton utility landing craft ordered from GRSE. The 63m long vessels can carry 145 tons of military equipment and 165 troops over a distance of around 1,500 nautical miles and attain a maximum speed of 15 knots. GRSE is also involved in the programme to build a total of sixteen so-called anti-submarine warfare shallow water craft, commencing work on its first vessel on 31 December 2020. Interestingly, the steel cutting ceremony was held at the private Larsen & Toubro facility at Kattupalli, near Chennai, which has been sub-contracted to build at least five of GRSE's eight contracted hulls. The other eight vessels will be built by CSL, which started production of its own lead unit earlier that month. Published information suggests that the two companies are producing somewhat different designs, although both will be armed with anti-submarine rockets and torpedoes and capable of maximum speeds in the region of 25 knots. CSL has also emerged as the lowest bidder for a contract for six next-generation missile vessels, somewhat larger ships that will replace some of India's existing aging missile-armed corvettes and fast attack craft.

Although previously reported as a confirmed deal, it seems that a contract for the construction of five, 45,000-ton fleet support ships to be built by Hindustan Shipyard with the assistance of the Turkish TAIS consortium has yet to be formally concluded. The deal has been controversial due to the Turkish shipbuilding sector's involvement with Pakistan navy programmes but the ships are urgently needed to supplement India's current force of replenishment vessels, which includes only two modern units.

OTHER INDIAN OCEAN NAVIES

Bangladesh: The Bangladesh Navy held a major ceremony on 5 November 2020 to commission five new vessels into its fleet. Three of these – the former People's Liberation Army Navy (PLAN) Type 053H3 'Jiangwei II' class frigates *Umar Farooq* and *Abu Ubaidah* and the newly built Chinese C-13B corvette *Prottasha* – had been delivered during 2019. However, two 33m *Darshak* class survey vessels built by the local Khulna Shipyard appear to be new arrivals. The Bangladesh Navy is increasingly emphasising local construction for smaller and less sophisticated ships as it seeks to expand its strength, with further 350-ton *Padma* class coastal patrol vessels and new survey boats also being built by Khulna. A tender for two larger, c. 700-ton patrol craft optimised for the anti-surface role was also floated early in 2020.

The navy suffered a major mishap on 4 August 2020 when the corvette *Bijoy* – the former British Royal Navy offshore patrol vessel *Dumbarton Castle* – was caught in the vast nitrate explosion that destroyed much of Beirut port. Twenty-one crew members were injured in the blast and the ship – part of the United Nation's UNIFIL mission – was sufficiently damaged to require towing to a Turkish port for temporary repairs. She eventually returned home in late October, being replaced by the new C-13B *Shongram*.

Myanmar: December 2020 saw the formal commissioning of *Minye Theinkhathu*, the Myanmar Navy's first submarine. The former Indian Navy Project 877EKM 'Kilo' class boat *Sindhuvir*, she had been delivered earlier in the year and participated in inaugural exercises in October. The same commissioning ceremony included the induction of two of a new type of 63m anti-submarine patrol vessel and a second 45m stealth fast attack craft, all indigenously

The Myanmar Navy's locally-built *Kyansitthar* class stealth frigates – the lead ship is seen here in September 2019 – are now being followed by construction of a larger frigate design. *(Myanmar Armed Forces)*

Pakistan's force of surface combatants – this is the Chinese-designed but KSEW-built F-22P frigate *Aslat* in June 2021 – is set to be reinforced by substantial deliveries of newly-built ships over the next few years. *(US Navy)*

built. Further units of both types seem likely as the navy attempts to replace 'legacy' littoral warfare ships that were delivered by China in the 1990s. A more significant development has been the resumption of frigate construction for the first time since the delivery of the second *Kyansitthar* class frigate *Sin Phyu Shin* in 2015. The new 135m vessel – Myanmar's fourth indigenous frigate – will be larger than her predecessors and is reported to be the first equipped with a vertical launch system (VLS). The Myanmar Navy has been increasing its reliance on Indian equipment in recent warships and it will be interesting to see if the relationship survives Myanmar's return to military rule and its subsequent crackdown on pro-democracy forces.

Pakistan: The last year has seen Pakistan continue to make good progress with its fleet renewal programme. This is being carried out in partnership with Chinese, Dutch and Turkish industry but also involves considerable input from the local Karachi Shipyards & Engineering Works (KSEW). Modernisation currently has four main strands, encompassing:

- Four Chinese Type 054A frigates: These are being built by the Hudong-Zhongua Shipbuilding at Shanghai, which commenced sea trials of the first vessel in May 2021. The second ship was launched on 29 January 2021 and the other two are at various stages of construction. The new ships are broadly similar to their Chinese counterparts but encompass differences in armament and electronic outfit.
- Four Turkish 'Milgem' light frigates: Construction of these vessels is being split between Istanbul Naval Shipyard and KSEW, which have commenced work on all four ships. The ships are believed to be somewhat modified from the Turkish 'Ada' variant and will reportedly carry the Albatros NG extended range variant of the CAMM surface-to-air missile system that is known as Sea Ceptor in British Royal Navy service. Deliveries are scheduled between 2023 and 2025.
- Two Damen *Yarmook* class corvettes: Built by Damen's Galati yard in Romania to the company's 'OPV1900' design, these ships were delivered without armament in February (*Yarmook*) and November (*Tabuk*) 2020 prior to final outfitting in Pakistan.
- Eight Chinese modified Type 039A 'Yuan' or *Hangor* class submarines: Just as for the 'Milgem' light frigates, construction of these boats will be split between the exporting nation and Pakistan, which is reportedly expanding shipbuilding facilities at KSEW for their construction.[8] Open source reports suggest that the first Chinese submarines will arrive before the end of 2022, with the overall programme set to be completed with the delivery of the final Karachi-built boat in 2028.

Taken together with the existing quartet of F-22P *Zulfiqar* class frigates and three 'Agosta 90B' type frigates, these should provide Pakistan with a fleet of around fourteen major combatants and eleven submarines by the end of the decade. According to remarks by Pakistan's retiring Chief of Naval Staff in October 2020, it is believed a force structure of some twenty large surface warships is ultimately envisaged. Other plans include more investment in fast attack craft and MCMVs and growth of the current small helicopter fleet.

Meanwhile, the imminent arrival of new warships is allowing the continued withdrawal of 'legacy' assets. Notably, the flotilla of former British Type 21 *Amazon* class frigates has been reduced to just two vessels following the disposal of *Shah Jahan* – the former *Active* – in a live firing exercise in January 2021.

Sri Lanka: A full review of the Sri Lanka Navy is contained in Chapter 2.3A.

AFRICAN NAVIES

A major step forward in **South Africa's** programme of fleet renewal occurred on 25 March 2021 when Damen launched the first of three 600-ton multi-mission inshore patrol vessels ordered under Project Biro. The new vessel is due to be delivered before the end of 2021 and will be followed by her two sisters at annual intervals. Work also continues on a new hydrographic survey vessel at Sandock Austral Shipyards to meet a scheduled 2023 delivery. However, a South African Department of Defence presentation to the parliamentary committee on Defence and Military Veterans has reported that funding remains inadequate to meet contractual obligations on the projects against a backdrop of continued reductions in the South African defence budget. This has meant money has had to be diverted from other priorities to sustain construction. A start has already been made on withdrawing the navy's remaining 'Warrior' (Modified Sa'ar 4) fast attack craft, which now operate in the patrol vessel role. *Galeshewe* was placed in reserve in October 2020 and will be followed by *Isaac Dyobha* in late 2021. This will leave *Makhanda* as the last of the class in service prior to her planned decommissioning at the start of 2023. Meanwhile, lack of funding is also preventing progress with already delayed mid-life modernisation of the four MEKO-200SA 'Valour' class frigates and three Type 209 'Heroine' class submarines that form the core of South Africa's naval combat capability.[9]

Elsewhere in Sub-Saharan Africa, a considerable amount of procurement of relatively unsophisticated but effective patrol craft to provide more effective policing of national waters is taking place against a backdrop of ongoing criminal activity. West Africa remains a particular hotspot of piracy. As many as 130 crew members – 95 percent of the global total – were kidnapped from vessels in the Gulf of Guinea in twenty-two separate incidents in the course of 2020.[10] **Nigeria** is at the centre of this dangerous area and has continued a programme of bolstering its constabulary assets. French company OCEA has been a major industrial partner over recent years, a relationship which has continued over the past twelve months with the delivery of two of the group's 35m FPB 110 series patrol vessels and four of its smaller 17m 'C-Falcon' interceptors. The navy has also continued to receive Suncraft 'Manta' fast patrol craft from Malaysia and is reportedly awaiting the arrival of additional vessels from Damen Vietnam and Aresa of Spain. Another major development has been the delivery of the OCEA-built hydrographic survey vessel *Lana*, which arrived in Lagos on 17 May 2021. A new tank landing ship, the Damen Sharjah-built *LST-1314*, was launched on 7 June 2021.[11]

In North Africa, **Egypt's** remarkable naval modernisation continues to dominate the headlines. Initially reported to be driven by a Saudi Arabian desire to create a 'proxy fleet' to help contain Iranian naval ambitions in the Red Sea, the fleet's expansion has been given additional impetus with the emergence of conflicting claims to economic resources in the Eastern Mediterranean. The most significant development has been the delivery of the Italian general-purpose FREMM frigate variants *Al-Galala* and *Bernees* in December 2020 and April 2021. Formerly destined for the Italian Navy as, respectively, *Spartaco Schergat* and *Emilio Bianchi*, their acquisition has allowed Egypt to commission new vessels without the delays normally associated with new construction. Also adding to the fleet has been the first indigenously-assembled 'Gowind 2500' type corvette, *Port Said*, which was commissioned on 6 January 2021. Joining the French-built *El Fateh* in service, she will be followed by two additional locally-built members of the class over the next couple of years. With the French variant FREMM *Tahya Misr* already in commission and four MEKO A-200 frigates on order from Germany, Egypt looks set to have a core fleet of eleven modern surface combatants – supplemented by older vessels – in service by the latter half of the current decade. Although substantial investment has been made in port facilities to support this new fleet, the maintenance of vessels from at least three different sources is likely to provide the Egyptian Navy with a substantial headache.

The Egyptian Navy is also upgrading other elements of its fleet. September 2020 saw the naming of *S44*, the last of four Type 209 submarines acquired from Germany's ThyssenKrupp Marine Systems (TKMS) under two contracts signed in 2011 and 2015. She is expected to be delivered before the end of 2021. German industry is also helping bolster coastal forces with the ongoing delivery of CSB-40 type patrol vessels from Lürssen. Nine of these patrol boats – and a larger 60m vessel – have been diverted from a larger order for the Saudi Arabian coast guard that was terminated by German government action after the 2018 murder of dissident Jamal Khashoggi in the Saudi consulate at Istanbul.

Further along the North African coastline, **Algeria's** previous acquisition spree has hit what is likely to be a temporary pause, with the only major unit commissioned over the last year being a second *El Kasseh* (improved *Lerici*) class minehunter from Italy's Intermarine. Delivered in July 2020, the ship is similar to Finland's *Katanpää* class and is reportedly part of a three-ship order. OCEA also continues to supply FPB-98 series patrol boats and is close to delivering the coastal survey vessel *El Masseh*. Elsewhere in North Africa, **Morocco** placed an order for what is likely to be an 'Avante' series offshore

The South African Navy's first new inshore patrol vessel – the as yet unnamed P1571 – pictured being transferred from the Damen Shipyards Cape Town facility to the Transnet National Ports Authority synchrolift at the Victoria & Alfred Waterfront Basin on the evening of 23 March 2021. *(Damen)*

Navantia is making good progress with building *Al Jubail* 'Avante 2200' class corvettes for the Royal Saudi Navy. This picture shows *Hail* – the third of the class to go down the slipway in the last year – being launched on 28 March 2021. *(Navantia)*

patrol vessel from Spain's Navantia. **Tunisia** is also bolstering its patrol forces in the light of the ongoing migrant crisis, taking delivery of the fourth of seven locally-built 28m *Istiklal* class patrol boats in April 2021.

MIDDLE EASTERN NAVIES

Although often associated with Egypt's naval expansion, **Saudi Arabia** is making progress with its own programme of maritime modernisation that is largely focused on the renewal of its Eastern, Persian Gulf-based Fleet. Two major procurement projects encompass four Lockheed Martin *Saud* class multi-mission surface combatants – derivatives of the *Freedom* (LCS-1) Littoral Combat Ship design – and five Navantia *Al Jubail* ('Avante 2200') corvettes. The latter contract is the more advanced, with three of the ships now in the course of fitting-out following the launches of *Al Jubail* in July 2020, *Al Diriyah* in November 2020 and *Hail* in March 2021 from the Spanish group's San Fernando yard in the Bay of Cádiz. The work on the Lockheed Martin vessels – just as for their Littoral Combat Ships – has been sub-contracted to Fincantieri's Marinette Marine yard, which laid the keel for the second of their quartet in late January 2021. Meanwhile a major milestone was reached on 14 October 2020 when Saudi Arabia's Zamil Offshore Company launched a French CMN-designed HSI132 interceptor, the first naval vessel ever built in a Saudi yard. The 32m vessel, reportedly capable of speeds in excess of 45 knots, is one of eighteen being built at the company's Dammam facility. A further twenty-one are being completed by CMN in Cherbourg, which commenced deliveries in mid-2019.[12]

Amongst Saudi Arabia's neighbours along the Persian Gulf, **Qatar** and the **United Arab Emirates (UAE)** also have significant naval procurement programmes underway. The former announced a massive contract with Fincantieri for a total of seven vessels – encompassing an amphibious transport dock, four air-defence 'corvettes' and two smaller but heavily armed 'offshore patrol vessels' – in June 2016. The first of the 3,250-ton corvettes, *Al Zubarah*, commenced sea trials in November 2020 and was followed by a second, *Damesh*, in July 2021. The lead offshore patrol vessel, the c. 750-ton *Musherib*, has also now ventured to sea, whilst her sister, *Sheraouh*, was launched on 5 June 2021. The programme marks a major expansion of the previously small Qatari Emiri Naval Forces, which is acquiring two specialised training ships from Turkey's Anadolu Shipyard to help support the resultant training burden. Both were launched in the second half of 2020 and will be delivered by 2022.

The UAE has also acquired Fincantieri-built vessels for previous modernisation projects but are increasingly turning to domestic construction to meet future requirements. In May 2021, they announced a c. US$1bn contract with local Abu Dhabi Shipbuilding (ADSB) to acquire four corvette-like 'Falaj 3' offshore patrol vessels in what was stated to be the company's largest-ever order. Singapore's ST Engineering has been contracted to provide design expertise in further evidence of the growing influence of Asian yards. The UAE are also acquiring two 'Gowind' corvettes from Naval Group. The lead ship should be launched before the end of 2021 and her sister-ship will follow in mid-2022. **Bahrain's** naval ambitions are more modest but a ceremony in February 2021 saw the formal commissioning of eight new patrol vessels of various sizes. Prominent amongst these was *Al Zubara*, the former British Royal Navy Falkland Islands' guard-ship *Clyde.*

The prime reason for naval expansion in the Gulf is the fear over **Iran's** intentions. The asymmetrically-focused Iranian Navy and separate Iranian Revolutionary Guard Corps (IRGC) Naval Forces have continued to harass 'Western' naval forces in the area as they strive for regional ascendancy as part of a campaign that has seen suspected clandestine attacks both by and against Iranian forces.[13] It has otherwise been a rather mixed year for the Iranian Navy, with the fleet's ability to operate beyond local waters significantly dented by the loss of its sole replenishment ship on 1 June 2021. The 33,500-ton *Kharg* – originally delivered by the British Swan Hunter yard in 1984 – went down off the Iranian port of Bandr-e-Jask in the Gulf of Oman after an engine room fire. Her sinking has been partly mitigated by the prior arrival of the multi-role support vessel *Makran*, a converted tanker that has similarities with the US Navy's expeditionary mobile base concept but which also has a replenishment capability. This was demonstrated when the ship supported a deployment of the frigate *Sahand* to the Atlantic in mid-2021. Other significant new arrivals include the fourth indigenous 'Mowj' class frigate – *Dena* – and *Shahin* – a new type of catamaran-hulled minesweeper – both commissioned at a ceremony at Bandar Abbas, the country's main naval base in the Persian Gulf, on 14 June 2021. The IRGC has also been taking delivery of additional ships, including their own floating base ship – *Shahed Roudaki* – and various types of fast interceptor.

Israel has also been receiving new vessels. The first of its TKMS-built 'Sa'ar 6' corvettes – *Magen* – was handed over at Kiel on 11 November 2020, being followed by *Oz* on 4 May 2021. A further two units currently under construction will be delivered before the end of 2021. However, the process of outfitting the new ships with their suite of largely indigenously manufactured weapons and electronic systems is being undertaken 'in country' by Israel Shipyards. As such, it will not be until 2022 that the new class starts entering operational service. Israel is also apparently awaiting the final hand over of its final 'Dolphin II' class submarine *Drakon.* The boat was apparently close to delivery early in 2020 when the COVID-19 pandemic disrupted final trials and crew training.

Notes

1. One recent example of changing allegiances has been conclusion of agreements to recognise Israel by the United Arab Emirates, Bahrain, Sudan and Morocco over the course of the past year.

2. The description of Libya as seeing 'the largest drone war in the world' is attributed to Ghassan Salamé, the former United Nations special envoy to the country. Chinese-made Wing Loong series drones supplied by the United Arab Emirates initially proved decisive in supporting the eastern-based Libyan National Army (LNA) faction make progress against the western territories of the then Government of National Accord headquartered in Tripoli. However, the arrival of Turkish forces – including Bayraktar TB2 and Anka UAVs – from early 2020 onwards first held back and then reversed the LNA's advance. A 'permanent' ceasefire between the rival factions was signed in October 2020, paving the way for the establishment of a Government of National Unity in March 2021.

3. For details of the factors driving Egypt's naval expansion and the strengths and weaknesses of the resulting force see the editor's 'Egypt: A Regional Naval Power?' published in *European Security & Defence 11-12/2020* (Bonn: Mittler Report Verlag, 2020), pp.46–50.

4. Political changes in Sudan have put the establishment of the new Russian base under question. As of mid-2021, the agreement allowing the facility was said to be under review against a backdrop of closer Sudanese ties with the United States.

5. A good review of Russia's strategy of expanding influence in the Middle East was provided by Sinan Ülgen and Can Kasapoğlu in a research paper 'Russia's Ambitious Military-Geostrategic Posture in the Mediterranean' posted to the *Carnegie Europe* website – carnegieeurope.eu – on 10 June 2021.

6. For further analysis of the budget, see Laxman K Behara's 'Firm on Modernisation: The Indian Defence Budget 2021-22' posted to the *Society for the Study of Peace and Conflict's* website – sspconline.org – on 10 February 2021.

7. See 'Explosion forced Indian Navy to return nuclear submarine to Russia?' posted to *The Week* website – theweek.in – on 9 June 2021.

8. See H I Sutton, 'Pakistan's New Type-039B AIP Submarines: Image Shows Shipyard Expansion' posted to the *Naval News* website – navalnews.com – on 6 October 2020.

9. The *defenceWeb* news portal – www.defenceweb.co.za – provides a unique source of ongoing coverage of defence news across the African continent. It contains much supplemental information to this chapter's summary.

10. Full details of global piracy are provided in the news section of the ICC's Commercial Crime Services website, accessible at: icc-ccs.org

11. For further detail on the Nigerian Navy's development see Mrityunjoy Mazumdar, 'The Nigerian Navy – making slow headway against heavy seas' *in Seaforth World Naval Review 2021* (Barnsley: Seaforth Publishing, 2020), pp.51–63.

12. See Jeremy Binnie, 'First Saudi-made naval vessel launched', *Jane's Defence Weekly* –21 October 2020 (Coulsdon: Jane's, 2020), p.13.

13. See another Jeremy Binnie article, 'Iran confirms attack on base ship in Red Sea', *Jane's Defence Weekly* –14 April 2021 (Coulsdon: Jane's, 2020), p.5. The article refers to a suspected Israeli limpet mine attack on the merchant ship *Saviz*, which has been claimed to be a base for Iranian operations supporting Houthi rebels in Yemen. The incident followed a number of other attacks on Iranian and Israeli merchant vessels.

2.3A FLEET REVIEW

THE SRI LANKA NAVY

A Navy in Transition

Author:
Mrityunjoy Mazumdar

The island nation of Sri Lanka – known officially as the Democratic Socialist Republic of Sri Lanka – lies at the southern tip of the Indian subcontinent. It is separated from India by fewer than 20 nautical miles at its narrowest point in the Palk Strait. With a coastline of some 1,340km, territorial waters of 21,500km^2, an exclusive economic zone (EEZ) of over 500,000km^2 and additional claims to the outer limits of its 350-nautical-mile continental shelf, Sri Lanka has significant maritime interests. Moreover, the country's proximity to trade and energy routes makes Sri Lanka an important strategic hub and, consequently, a focal point for the emerging naval rivalry between India and China in the Indian Ocean Region (IOR).

Granted independence from the British Empire in 1948, what was previously Ceylon became a republic within the Commonwealth and adopted its current name in 1972. Periods of instability were most prominently marked by the on/off c. 25-year civil war between 1983 and 2009 against Sri Lankan Tamil separatists – the Liberation Tigers of Tamil Eelam (LTTE) – fighting for an independent homeland. At times, the LTTE had a significant presence along around a third of Sri Lanka's coastline and across a quarter of its landmass. This – and other insurgencies – left perhaps some 200,000 dead against a backdrop of atrocities perpetrated by all sides. Sri Lanka's ultimate victory in this conflict owed much to the build-up of naval capabilities to interdict LTTE supplies and deal with the threat posed by its maritime arm, known as the Sea Tigers.

With the end of the so-called Eelam Wars and a significant diminution of the threat from maritime terrorism, other age-old maritime security problems caused by non-state actors – typified by smuggling, people trafficking, illegal fishing and pollution – came back into focus. Given that world opinion and adroit diplomacy should keep Sri Lanka safe from any threat from its neighbours, these are currently the main issues faced by the Sri Lankan sea services.

Two indigenously-built 'Colombo Mk IV' fast attack craft – *P436* (foreground) and *P435* – of the Sri Lanka Navy pictured in 2019. The Sri Lanka Navy is slowly transitioning from a green and brown water counter-insurgency force to a more balanced fleet better able to perform constabulary duties into the vast expanses of the Indian Ocean. *(Tetsuya Kakitani)*

HISTORIC DEVELOPMENT

Like most Commonwealth navies, the Sri Lanka Navy (SLN) traces its roots to the British Royal Navy. In 1937, it was decided to establish a Ceylon Naval Volunteer Force (CNVF) as part of moves to strengthen the then British Empire's local maritime defence forces. Subsequently absorbed into the Royal Navy as the Ceylon Royal Naval Volunteer Reserve (CRNVR) in 1942, this fledgling force operated an assortment of tugs, trawlers and miscel-

THE SLN AND THE EELAM WAR (1983–2009)

The Sri Lankan Civil War was fought in four main phases – interspersed with ceasefires and failed peace talks – between 1983 and 2009. During this time Sri Lankan Tamil separatists of the LTTE based largely in the north and east of the country fought a bitter civil war for their own homeland. Amongst their many forces, the LTTE developed a potent, asymmetric naval arm known as the Sea Tigers, which included a suicide cadre known as the Black Sea Tigers.

Numbering perhaps 3,000 men and women, the Sea Tigers were arguably the best-organised and most effective non-state navy of recent times. Availing themselves of excellent intelligence, their asymmetrical warfare tactics – including the use of attack and suicide craft, mining operations and underwater demolition teams – posed a severe threat to SLN operations. During combat at sea, the Sea Tigers displayed remarkable levels of tactical acumen and aggression. They cleverly exploited local and Indian fishing fleets to their advantage by hiding their stealthy, heavily-armed interceptors and suicide craft amongst civilian fishing vessels. Then they mounted surprise swarm attacks on lone or small groups of SLN patrol craft that were largely unable to cope with the overwhelming nature of the assault. SLN casualties were high and – at their peak – the Sea Tigers were able to exercise effective sea denial in various locations and at various times to Sri Lankan forces.

The LTTE operated a sophisticated international logistic network of merchant ships that included floating armouries and vessels – known as Sea Pigeons – that were used for transporting weapons, war materiel and personnel into Sri Lanka. Flying flags of convenience and often bearing fictitious names, this fleet engaged in legal activities when not engaged in transporting arms and supplies, thus making them quite difficult to track. Typically, the larger vessels carrying war materiel would stay well offshore when undertaking supply missions, offloading their cargoes onto smaller vessels which would be escorted by Sea Tiger gunboats into Sri Lankan waters.

Although hard pressed at first, the SLN slowly effected doctrinal changes to warfighting based on an appreciation that Sea Tiger swarm tactics could only be fended off by heavy firepower. Tactically, the SLN adopted Sea Tiger tactics and bettered them, eventually overwhelming the Sea Tigers with large numbers of heavily-armed craft deployed in conjunction with the use of land-based radar sites to detect and disrupt LTTE activities. Up-gunned fast attack and smaller patrol craft commanded by a new generation of highly motivated and competent young officers proved much more successful in combatting the Sea Tigers than in the early stages of the fighting.

Strategically, it took longer for the SLN to realise that the key to winning the war lay in choking off the LTTE's sea-based logistics. After acquiring suitable long-range OPVs, the SLN went after the LTTE's floating warehouses in the later stages of the war, using a combination of intelligence inputs from its own and allied sources. By 2009, all the known LTTE floating warehouses had been eliminated. Meanwhile, an ever-tightening sea blockade of rebel-held areas increasingly impacted the LTTE's room for manoeuvre.

Ultimately, therefore, the SLN was able to exert a significant strategic influence on the war, decisively turning the tide on the Sea Tigers and crippling broader LTTE operations by interdicting its logistical supplies. The cost was heavy. The navy had lost some 400 personnel in naval engagements, with eight major vessels, 20 fast attack craft and 28 smaller patrol craft destroyed or captured in these actions, by the time the war was over.[1]

1. See *Humanitarian Operation Factual Analysis July 2006-May 2009* (Colombo: Ministry of Defence – Democratic Socialist Republic of Sri Lanka, 2011), p.19. The document can be downloaded at: slembassyusa.org/wp-content/uploads/2011/08/Sri-Lankan-Humanitarian-Operation-Factual-Analysis.pdf

The 'Shanghai II' fast gunboat *Weeraya* was one of many examples of the type that Sri Lanka acquired from China from the early 1970s onwards. They proved largely inadequate to counter the threat from the LTTE's Sea Tigers in the earlier phases of Sri Lanka's civil war. (*Sri Lanka Navy*)

laneous auxiliary vessels in local defence, search and rescue and escort roles throughout the conflict. Reverting to Ceylonese control at the end of the war, the CRNVR would go on to serve as the nucleus of an independent Ceylonese Navy after independence. The new Royal Ceylon Navy was formally established after the approval of the necessary legislation on 9 December 1950.

Initially still very much under the shadow of a continuing Royal Navy presence, the navy's profile was considerably strengthened after the final withdrawal of British forces from the island in 1956 and the acquisition of the former HMS *Highflyer*, the major naval base located at Trincomalee. The navy also purchased a number of larger vessels, most notably two Canadian-built 'River' class frigates obtained from Israel that were delivered during 1959–60. However, any 'blue water' aspirations the navy might have held proved to be short lived, with political change being accompanied by a shift towards smaller patrol craft and an emphasis on the constabulary security of the littorals as part of an overall de-emphasising of the navy's importance. The Marxist Janatha Vimukthi Peramuna (JVP) insurrection of 1971 brought a foretaste of the internal conflict that was to be a feature of subsequent decades and was followed by a period of significant expansion of what, after the formation of the republic in 1972, became the SLN. Amongst notable acquisitions were the first of several large, Chinese built 'Shanghai II' series patrol vessels.

The outbreak of the Eelam Wars in 1983 further increased the navy's importance. Important acquisitions included what would ultimately be a large fleet of Israeli-built Dvora fast attack craft (FAC) along with converted merchant ships for use as mobile surveillance command ships to control the smaller vessels. During this time, the SLN mounted its first amphibious operation, added naval infantry forces for landward counter-insurgency operations and created a Special Forces unit – the Special Boat Squadron (SBS). Combat losses proved to be heavy, with the navy's then head – Admiral W W E Clancy Fernando – being assassinated by a LTTE suicide bombing in 1992. Ultimately a fundamental shift in strategy and doctrine saw the SLN focus on destroying the LTTE's sea-based logistical supply train, which had a major impact on ending the war. Since the end of hostilities, the SLN has effectively been undergoing a period of transition from a wartime counter-insurgency force to a peacetime navy with expanded constabulary capabilities.[1]

The Sri Lanka Navy's Indian-built offshore patrol vessels *Sayura* (left) and *Sagara* pictured on a port visit to the Indian Navy's main eastern coast base at Visakhapatnam in September 2017. The SLN maintains close naval ties with other regional fleets as it attempts to extend its reach with larger vessels but lack of resources and the legacy of a force architecture designed for counter-insurgency warfare remain important challenges. *(Indian Navy)*

CURRENT NAVY ROLES AND RESOURCES

The SLN currently regards itself as having three main roles, viz. (i) military, (ii) diplomatic, and (iii) constabulary. The current focus is very much towards constabulary duties, including the surveillance and protection of both territorial seas and the oceanic EEZ. In this role, the navy is supported by the Sri Lanka Coast Guard (SLCG), which was established in its current form in 2009 and carries out policing functions under the control of Sri Lanka's Ministry of Defence. Another important relationship is with the Sri Lanka Air Force, particularly through the naval air surveillance unit.

Sri Lanka maintains a broad array of bilateral and multilateral maritime security cooperation arrangements with Indian Ocean rim countries and beyond, as well as membership of numerous regional and extra-regional security cooperation frameworks. The SLN maintains close naval ties with other fleets in the region, as well as the global sea powers – particularly the United States and China – that are active in the region. Many of these fleets have provided support to the SLN in terms of equipment, training and other assistance. The SLN also has much to offer in return, being one of a few navies to have fought – and won – an intensive counter-insurgency based naval war.

The navy produced a strategic vision – *Sri Lanka Navy's Maritime Strategy 2025* – for limited consumption in 2016. An updated version of this document was published – the first time such a doctrine had been released to the Sri Lankan public – as the *Maritime Doctrine of Sri Lanka* (MDSL) in July 2020.[2] It outlines the broad contours of maritime power's employment under the strapline of 'nurturing a stable environment at sea'. Its essence envisages the SLN's transition from a coastal force with limited offshore capabilities to a fleet that is more capable of projecting its influence across Sri Lanka's extensive interests in the Indian Ocean. In addition to requiring more capable vessels, this strategy is seen as being reliant on the improved surveillance capabilities provided by networked technology and a greater focus on aerial capacity, both manned and unmanned. Other areas of focus include an emphasis on training and professionalism, as well as improved inter-operability with sister services and regional partners. Clearly achievement of this vision will require considerable investment, a challenge which current spending is not best structured to meet.

Sri Lanka's defence budget is currently c. US$1.7bn p.a., of which around US$300m is allocated to the navy. Overall defence spending and the navy's budget share have both fluctuated in recent years, with the latter being heavily impacted by capital expenditure requirements. For example, an uptick in the middle of the last decade corresponded with new offshore patrol vessel procurement. In common with many developing navies, there is an ongoing struggle to direct adequate funds towards investment in equipment, with capital expenditure averaging only a little more than 10 percent of naval spending over time. SLN manpower grew rapidly during the later stages of the Eelam Wars but has not returned to previous levels well over a decade since the civil war's end in spite of calls from within the navy to 'right-size' the force in terms of personnel. As a result, spending continues to be dominated by servicemen's 'personal emoluments' at the expense of new ships and weaponry.

NAVY SIZE AND STRUCTURE

Whilst precise data on SLN personnel numbers is difficult to ascertain, it seems current numerical strength is c. 55,000, including over 3,000 officers. These figures include officers and sailors on secondment to the SLCG Perhaps as little as 10 percent of this strength occupy seagoing billets, with the rest involved in administrative, engineering and technical roles or serving as naval patrolmen (infantry) in a legacy from the civil war.

The Commander of the Navy, currently Vice Admiral Nishantha Ulugetenne, exercises operational and administrative control of the navy from SLN Naval Headquarters in Colombo. He is assisted by a Chief of Staff and a Deputy Chief of Staff, through whom various specialised directorates report as part of the SLN Board of Management. He also has a considerable number of direct reports, the most importance of which are the Naval Area Commands and the Naval Fleet Command, the latter being sub-divided into various squadrons for administrative purposes.

As of mid-2021, the SLN operates seven Naval Areas or Naval Commands, the two terms apparently being used interchangeably. These are the Northern Command (HQ: Kankesanthurai); North Central Command (Thalaimanar); North Western Command (Mullikulama); Western Command (Colombo); Southern Command (Galle); Eastern Command (Trincomalee); and South Eastern Command (Panama). Each of these commands is headed by a flag officer and has a headquarters base as well as, typically, several additional bases and smaller naval detachments.[3] These include many inland outposts for naval infantry. The principal naval base is at Trincomalee on the eastern seaboard. The former British Royal Naval Dockyard, the now Sri Lanka Naval Dockyard spreads over 850 acres and has access to one of the finest natural harbours in the world. It is home to the Naval & Maritime Academy (NMA), headquarters of the Eastern Command and, *inter alia*, the home port of the navy's fast attack flotilla. Other important base facilities include Sri Lanka Navy Ship (SLNS) Rangalla in the Port of Colombo and the recently commissioned SLNS Kawanthissa at the Chinese controlled port of Hambantota. It remains to be seen whether the previously-planned transfer of the Southern Command headquarters from Galle to Hambantota will materialise.

The AOPVs *Gajabahu* – the former US Coast Guard high endurance cutter *Sherman* (WHEC-720) – and *Sayurala* – a newly-built *Saryu* class ship – pictured exercising with an unidentified fast attack craft and a US Navy helicopter in June 2021. The Sri Lanka Navy's fleet of seven OPVs are arguably its most important assets as it attempts to expand the scope of its constabulary missions. *(US Navy)*

NAVAL VESSELS

There are currently c. 700 vessels and craft of various sizes in SLN service, the vast majority of these being 'brown water' fibreglass boats and dinghies. Very few are capable of venturing safely beyond coastal waters, with EEZ surveillance under the responsibility of just a handful of larger offshore patrol vessels (OPVs). Force structure is summarised in Table 2.2A.1 with details of principal units following.

Offshore Patrol Vessels: The SLN's most important warships are arguably its seven gun-armed OPVs – all capable of helicopter operation – of various vintages and origins. The four largest are designated advanced offshore patrol vessels (AOPVs). Along with the smaller, locally-built 39m patrol vessel *Jayasagara*, these ships form the core of the SLN's ability to operate beyond the immediate littoral zone.

The largest ship in the SLN is the former United States Coast Guard *Hamilton* class high endurance cutter *Gajabahu*, formerly *Sherman* (WHEC-720). Originally built in the late 1960s, the c. 3,300-ton vessel was acquired in 2018 and commissioned the following year. Her principal armament is a single 76mm Oto Melara gun. The SLN hopes to acquire a second high endurance cutter in the near future.

Reflecting the various nations vying for influence in the Indian Ocean region, the SLN also commissioned the former Chinese Type 053H2G frigate *Tongling* in 2019. Stripped of its missile armament and converted for constabulary duties, the c. 2,500-ton, 1990s-built ship has been named *Parakramabahu* in Sri Lankan service. Armament includes a twin 100mm main gun and four twin 30mm secondary mounts. There have been reports that the ship has suffered technical problems whilst operating with the SLN and her sea-time has been limited.

The bulk of the OPV forces comprise ships of Indian origin. Prominent amongst these are two *Saryu* class AOPVs ordered from Goa Shipyard

Limited (GSL) in early 2014 and which entered service as *Sayurala* and *Sindurala* in, respectively, 2017 and 2018. A little over 105m long and displacing around 2,300 tons, the ships are capable, long-ranged vessels but currently equipped with only light-calibre weapons. The first unit reportedly cost some US$67m at then current exchange rates, representing a significant success for the Indian shipbuilding sector.

The other two Indian-built OPVs are the 1,900-ton *Sarayu* – a former Indian Navy *Sukanya* class vessel acquired second-hand in 2000 – and the smaller 1,200-ton *Sagara* – previously an Indian Coast Guard *Vikram* class OPV – that entered service in 2006. Along with the former US Coast Guard *Reliance* class medium endurance cutter *Samudura* – previously *Courageous* (WMEC-622) – commissioned in 2005, these smaller OPVs were acquired in the later stages of the Eelam Wars to bolster the SLN's ability to conduct long range interdiction missions against LTTE supply ships.[4]

Officially classified as an OPV, the Colombo Dockyard-built *Jayasagara* is a smaller vessel commissioned in 1983 and without the aviation facilities of the other ships. Her sister, *Sagarawardena*, was sunk in an attack by LTTE explosive suicide boats in September 1994.

Patrol Vessels: There are between ten and fifteen medium-sized patrol vessels currently in SLN service. The two largest and most powerful are the two former Israeli 58m Sa'ar 4 class 'fast missile vessels' *Nandimithra* and *Suranimala*. Acquired in 2000, the c. 400-ton ships are nominally fitted with Gabriel missile launchers. However, it is doubtful if the missiles were ever used operationally by the SLN and, in practice, the launchers have been replaced by stabilised small-calibre gun mountings.

Two former Austal-built, Australian Customers Service 'Bay' class coastal patrol vessels were commissioned in 2014. Donated with a view to improving the SLN's capacity to stem people smuggling to Australia, the 38m, c. 135-ton *Mihikatha* and *Rathnadeepa* are armed only with light machine guns.

The balance of the large patrol vessel force comprises a number of fast gunboats of Chinese origin. Comprising Type 062 'Shanghai II' (P31 series); 'Modified Shanghai II' (P32 series); Type 062/1G 'Haizhui' (P33 series) and 'Modified Haizhui'/'Lushun' (P34 series) types, these are 39m to 46m vessels with displacements ranging from 140

Based on the Indian Navy's *Saryu* class, *Sindurala* is the second of two AOPVs ordered from Goa Shipyard in 2014 and which were delivered in 2017 and 2018. The ships are lightly armed with largely obsolescent weaponry but significantly expand the SLN's oceanic reach. *(Chris Sattler)*

Sagara – previously the Indian Coast Guard *Vikram* class OPV *Varaha* – was acquired on free loan during the later stages of Sri Lanka's civil war to improve the SLN's ability to intercept LTTE maritime logistics. It was subsequently gifted to the SLN by India in 2015. A sister-ship is operated by the Sri Lanka Coast Guard. *(Tetsuya Kakitani)*

Table 2.3A.1: SRI LANKA NAVY – PRINCIPAL UNITS AS AT MID-2021[1]

TYPE	CLASS	NO.	YEAR[2]	TONNAGE	DIMENSIONS	SPEED	ORIGIN
Offshore Patrol Vessels (7)							
Patrol Vessel (OPV)	SAYURALA	2	2017	2,300 tons	105m x 13m x 4m	25 knots	India
Patrol Vessel (OPV)	Type 053H2G ('Jiangwei I')	1	1992	2,500 tons	112m x 12m x 5m	27 knots	China
Patrol Vessel (OPV)	SUKANYA	1	1989	1,900 tons	101m x 12m x 4m	21 knots	India
Patrol Vessel (OPV)	VIKRAM	1	1983	1,200 tons	74m x 11m x 3m	22 knots	India
Patrol Vessel (OPV)	HAMILTON (WHEC-715)	1	1967	3,300 tons	115m x 13m x 5m	29 knots	United States
Patrol Vessel (OPV)	RELIANCE (WMEC-615)	1	1964	1,100 tons	64m x 10m x 3m	18 knots	United States
Another *Vikram* class ship serves in the Sri Lanka Coast Guard							
Patrol Vessels and Gunboats (c. 14)							
Coastal Patrol Vessel (CPV)	'Bay'	2	1999	135 tons	38m x 7m x 2m	24 knots	Australia
Fast Gun Boat (PV)	Type 062/1G (Mod 'Haizuhi')	2	1992	215 tons	46m x 6m x 2m	28 knots	China
Fast Gun Boat (PV)	Type 062/IG ('Haizuhi')	c. 3	1992	175 tons	41m x 5m x 2m	25 knots	China
Patrol Vessel (PV)	JAYASAGARA	1	1983	350 tons	40m x 7m x 2m	15 knots	Sri Lanka
Fast Patrol Vessel (PV)	SA'AR 4	2	1973	450 tons	58m x 8m x 2m	35 knots	Israel
Fast Gun Boat (PV)	Type 062 (Mod 'Shanghai II')	1	1961	150 tons	41m x 5m x 2m	29 knots	China
Fast Gun Boat (PV)	Type 062 ('Shanghai II')	c. 3	1961	140 tons	39m x 5m x 2m	30 knots	China
The SLN only lists five Chinese fast gunboats in service and officially announced the retirement of the old Type 062s *Weeraya* and *Jagatha* in 2020. An additional two Modified Type 062s and two Type 062/IGs were noted as operational in the course of 2020 and are included in the above list.							
Amphibious Forces (5)							
Landing Ship (LSM)	Type 074 ('Yuhai')	1	1995	800 tons	58m x 10m x 3m	14 knots	China
Landing Craft (LCM)	RANAGAJA	2	1983	270 tons	33m x 8m x 2m	8 knots	Sri Lanka
Landing Craft (LCM)	Type 067 ('Yunnan')	2	1968	140 tons	29m x 5m x 2m	12 knots	China
The SLN also operates a self-propelled barge, two small fast utility landing craft and three fast personnel carriers.							
Other Vessels							
The core of the navy is formed of its flotilla of heavily-armed fast attack craft. This comprises around fifty Israeli 'Shaldag' and 'Super Dvora', US 'Trinity Marine' and indigenous 'Colombo' types of various iterations and which displace between 50 and 75 tons. There are also over 100 'Wave Rider' and 'IPC-72' series inshore patrol craft of c. 5 to 10 tons displacement along with approximately 160 smaller 'Cedric' type interceptors. Other vessels include an auxiliary trawler, some smaller fishing vessels, a handful of tugs and around 300 fibreglass dinghies. The SLN Hydrographic Service also operates RV *Samudrika*, a research vessel from Sri Lanka's National Aquatic Resources Research and Development Agency (NARA) since February 2019.							

Notes

1. This table is based on publicly available sources, some of which are contradictory. Some information should therefore be regarded as speculative.

2. Date relates to the date the first ship of the class entered service, sometimes with a foreign navy. In some cases, these ships are now retired.

The Sa'ar 4 fast missile vessel *Nandimithra* pictured outboard of the OPV *Samudura*. The SLN's two Sa'ar type fast patrol vessels are nominally its most potent assets but their surface-to-surface missiles have long been dismounted. *(Indian Navy)*

to 215 tons. Eleven of the class were in service at the turn of the decade but numbers are starting to decline as older units start to be retired, with probably around nine in active service as of mid-2021.

Fast Attack Craft: The 'heart' of the SLN's fleet comprises around fifty fast attack craft (FACs) of the Trincomalee-based 4th Fast Attack Flotilla (FAF4), formerly the 4th FAC Squadron. Though the bulk of the FACs are based at Trincomalee, detachments deploy to the other principal naval bases.

The SLN's FAC fleet expanded rapidly from the mid-1980s onwards as a counter to the LTTE's Sea Tigers. The first of many Israeli 'Dvora' and 'Super Dvora' class FACs entered service in 1984, later being supplemented by Israeli 'Shaldag' and US 'Trinity Marine' types. In addition to these imported vessels, over thirty indigenous FACs were constructed by Colombo Dockyard to progressively modified 'Dvora' designs from the 1990s onwards. FACs, regardless of the type, are colloquially known as 'Dvoras' in the SLN.

Initially, the SLN's FAC tactics were hampered by a lack of suitable fire-control optics for night-time operations and inadequate firepower. This led to the loss of several FACs that were overwhelmed by Sea Tiger swarm attacks and suicide craft. Once the SLN's leadership realised that the Sea Tiger swarm tactics could only be fended off by heavy firepower, the SLN up-gunned their FACs and adopted a more aggressive approach to warfighting that ultimately prevailed, albeit at the expense of continuing losses. Given the high tempo of wartime FAC operations, the SLN introduced an 'additional crew' concept to relieve tired and exhausted personnel.

Today, the FACs are actively involved in coastal constabulary missions to eliminate a wide range of criminal activity, as well as additional duties such as search and rescue. Efforts continue to be made to enhance the existing craft, for example through ongoing upgrades to sensors. Current inventory – all with P4xx/P4xxx numbers – comprises approximately five 'Shaldag', twenty-five 'Colombo', fifteen 'Dvora' and 'Super Dvora', and five 'Trinity Marine' FACs. A handful of FACs – primarily 'Dvora' type vessels and South Korean-built 'PKM' class vessels – have been transferred to SLCG service.

Inshore Patrol Craft and Combat Boats: The SLN's small inshore patrol craft can – in very broad terms – be split between its 'Waver Rider' and 'IPC' types, the latter encompassing a number of designs with broadly similar characteristics but numerous differences in size and detail. With lengths ranging from 13m to 15m, all are heavily-armed vessels equipped with inboard diesels and water jets and capable of 25–35 knots. Very large numbers of these craft were built in the course of the Civil War, with all construction since 2000 taking place at the Naval Boat Building Yard, SLNS Thakshila, at Welisara north of Colombo.

Supplementing these inshore types are the even

The SLN continues to operate a number of Chinese-built fast gunboats of the 'Shanghai II' and subsequent series but numbers are dropping as older vessels reach retirement age. This is the Type 062/1G 'Lushun' class *Prathapa*, first delivered in the late 1990s. *(Sri Lanka Navy)*

A photograph of the Sri Lanka Navy 'Super Dvora' series fast attack craft *P4442* taken shortly after the end of the civil war at a fleet review in December 2010 to mark the SLN's 60th anniversary. In the background are naval vessels that include a former LTTE merchant ship *We Ling*. The image clearly indicates the efforts made by the SLN to upgrade their vessels with improved armament and sensors to combat Sea Tiger swarm attacks *(Sri Lanka Navy)*

The Sri Lanka Navy built large numbers of small inshore patrol craft of various iterations during the course of the civil war. *P126* – built by Blue Star Marine in Colombo in the late 1990s – is typical of these types of vessel. *(Tetsuya Kakitani)*

The Type 074 'Yuhai' class landing ship *Shakthi*, acquired from China in 1994, is the Sri Lanka Navy's most significant amphibious asset. She is capable of transporting up to 150 tons of equipment and supplies over distances of up to 1,000 nautical miles. *(US Navy)*

more numerous 'Cedric' ('Arrow') type interceptors.[5] Fitted with inboard diesels and water jets, these are smaller, c. 6–7m combat boats capable of up to 40 knots but more lightly armed than the inshore patrol craft. Their construction has also been focused on the Naval Boat Building Yard. A combined total of a little over 100 'Wave Riders' and 'IPCs' remain in the navy's inventory, along with around 160 'Cedrics'. Many are in use with the Special Boat Squadron and Rapid Action Boat Squadron, described further below, with a few 'IPCs' also in coast guard service. It is also possible that a handful of larger but typically slower coastal patrol craft remain operational, some with the SLCG.

Amphibious Vessels and Auxiliaries: The SLN's small flotilla of amphibious warships provided a vital lifeline for beleaguered government forces during the Eelam War, particularly for the supply of troops at Kankesanthurai on the Jaffna peninsula after the LTTE managed to cut off all land access. The largest vessel currently in service is the 800-ton Chinese-built Type 074 'Yuhai' class landing ship *Shakthi*, acquired in 1994. She is supplemented by two smaller 300-ton *Ranagaja* class landing craft built at Colombo Dockyard and two 150-ton Chinese Type 067 'Yunnan' craft. *L810*, a self-propelled steel barge, also provides a modest additional amphibious transport capability. The navy also operates two new fast utility landing craft. *L801* has clamshell doors and a small bow ramp for discharging troops while *L802* has a bow ramp and deck space for a light pickup truck.

Other SLN ships include three catamarans used as fast troop transports as well as the large converted trawler *A521*, now in service as a diving support and patrol vessel. There are additional small fishing vessels, tugs and yard craft, including as many as 300 fibreglass dinghies used for various patrol and auxiliary duties.

NAVAL INFANTRY AND SPECIAL FORCES

As previously mentioned, the SLN's activities extend far beyond the open seas, including substantial numbers of infantry, marine and Special Forces elements. Some of the more important of these units are discussed further below.

Naval Patrolmen: The SLN has long used naval personnel for shore patrol and internal security duties. During the Eelam Wars, the expanded need for base and port security led to the creation of the Naval Patrolmen (NPM) Branch. The branch's remit steadily expanded from base protection to broader combat operations as a nominally amphibious infantry force equipped with light armoured vehicles. This shift was driven by the need to bolster the Sri Lanka Army after it had suffered several reverses at the hands of the LTTE. The branch was used extensively in combat operations and wartime casualties were significant.

The NPM Branch is organised into battalions and located at the various naval bases and outposts. Organisationally, the branch comes under the command of the Director Naval Land Operations & Commandant Naval Infantry, who in turn reports to the Director General Operations (DGO) on the SLN Board of Management. Operationally, units are under the control of the Naval Command where they are located. In total, it is believed there are c. 5,000 patrolmen currently in service.

Modelled after the British Special Boat Service and the US Navy SEALS, Sri Lanka's Special Boat Squadron introduced 'Cedric' (formerly 'Arrow') type craft such as the one pictured that were used to great effect whilst undertaking a range of special operations missions against the LTTE during the civil war. *(Tetsuya Kakitani)*

Sri Lanka Marine Corps: The rationale for maintaining such a large naval infantry force – often not easily identified as being focused on performing maritime tasks – after the end of the civil war is debateable and there have been calls from within the navy for it to be downsized. The formation of the Sri Lanka Marine Corps (SLMC) in 2016 using a core of former naval patrolmen may be part of this process. The initial goal was to create a two-battalion force comprising the 1st and 2nd Marine Battalions of c. 600 personnel each. Headquarters and training facilities are at SLNS Vidura, Sampoor near Trincomalee and, just as for the NPM Branch, reporting is through the Director Naval Land Operations & Commandant Naval Infantry. Marines have already been deployed on flood rescue missions and for riot control duties. Future plans include inland water operations and training for UN peacekeeping missions.

Both Sri Lanka's Special Boat Squadron and its Rapid Action Boat Squadron, which also forms part of the Maritime Special Force directorate, continue to operate large numbers of inshore patrol craft and similar vessels in both security and natural disaster response roles. (US Navy) *(US Navy)*

Special Forces: Modelled after the British Special Boat Service and the US Navy SEALS, Sri Lanka's Special Boat Squadron (SBS) pre-dates the SLMC, being established during the civil war by then Lt. Commander R C Wijegunaratne, who would later rise to become Chief of Defence Staff of the Sri Lanka Armed Forces.[6] The SBS introduced 'Arrow' (since renamed as 'Cedric') type craft that were used to great effect whilst undertaking a range of special operations missions against the LTTE during the civil war and their boarding teams have continued relevance as the SLN transitions towards a more constabulary-focused model. The SBS reports through the Director Maritime Special Force to the DGO.

Sometimes associated with SBS operations and also part of the Maritime Special Force directorate is the SLN's Rapid Action Boat Squadron (RABS). This specialises in small-boat operations in support of amphibious and riverine missions. Another child of the civil war, this was created to counter Sea Tiger swarm boat tactics through the use of inshore patrol craft that were more agile than previous vessels. Its duties now encompass a wider range of emergency response operations, as evidenced by the establishment of a Rapid Response Rescue and Relief Squadron using RABS sailors. The squadron is believed to have in the region of 1,000 personnel including five RABS teams focused on naval missions and five rescue and relief teams of life-saving qualified sailors.

The SLN's mine clearance and salvage diving teams played an important role during and after the civil war in dealing with improvised mines and other obstructions. Like other elements of the Special Forces, they have continued relevance responding to natural disasters.

NAVAL AVIATION

The SLN lacks an organic air arm so all air maritime operations – rotary and fixed wing – are undertaken by the Sri Lanka Air Force (SLAF). The SLAF's maritime reconnaissance unit is No. 3 Maritime Squadron located at China Bay, Trincomalee. It is equipped with two Beechcraft King Air B200 and some Harbin Y-12 aircraft. SLN 'maritime-air-reconnaissance officers' fly aboard these aircraft. The SLAF is expected to take delivery of two additional King Air 360ER Extended Range (B300) maritime patrol aircraft in the near future. The SLAF also operates two squadrons equipped with unmanned aerial vehicles (UAVs) and data they obtained can also be integrated into the joint Air and Maritime Strategic Command Centre located at SLAF headquarters.

Sri Lanka Navy clearance and salvage divers seen in the course of an exercise with their US Navy counterparts at the vast Trincomalee base. The navy's divers played an important role in the civil war and continued to be involved in clearing up the unexploded ordnance left in its aftermath. *(US Navy)*

Although the SLN's OPVs are equipped for helicopter operations, there is little evidence that these are regularly embarked. However, 'at sea' landing trials utilising one of the SLAF's Bell 412 helicopters were carried out aboard *Sindurala* in late 2019. Since then, helicopters have been deployed from OPVs during naval exercises.

MARITIME SURVEILLANCE

Sri Lankan maritime domain awareness was relatively limited in the early stages of the civil war but has improved significantly during the current millennium. As of today, the coastal surveillance system is believed to encompass over seventy coastal radar stations largely using commercial Furuno E/F band (US S band) and I/J band (US X band) sets, a Raytheon Canada HF-SWR-503 high frequency surface wave radar with a 200-nautical-mile range and around 250 coastal observation positions with electro-optical surveillance systems. The network supports automatic identification system (AIS) recognition, with data fed into the Air and Maritime Strategic Command Centre and other relevant headquarters. The sites are heavily concentrated in the north and east (areas formerly under LTTE control) and around the capital port of Colombo but gaps in the network are being steadily filled.

Given only limited numbers of long-range OPVs and maritime patrol aircraft, surveillance is likely to be more limited in the areas of Sri Lanka's EEZ that extend beyond the coastal belt. SLN officers have also admitted that there are serious gaps in the navy's ability to generate a sub-surface picture. Efforts are being made to remedy some of these deficiencies, for example through the deployment of satellite-based AIS. Another development is growing efforts towards international cooperation, with Indian media reports suggesting that India is helping to establish a more powerful chain of coastal surveillance radars as part of its efforts to improve overall maritime domain awareness – and counter Chinese influence – across the Indian Ocean region.

NAVAL SHIPBUILDING AND RESEARCH

Although the SLN has been reliant on overseas suppliers for its major warships, it is largely self-sufficient in smaller craft. The premier indigenous builder of fast attack craft is Colombo Dockyard, which first built patrol vessels in the 1970s and progressed to volume production of FACs from the mid-1990s onwards. It may well be involved in fulfilling the navy's future expansion plans. The other major supplier is the SLN's own Naval Boat Building Yard, which focuses on smaller inshore patrol craft and 'Cedric' craft. Both yards have had export success, with the Naval Boat Building Yard gaining a substantial US$4.2m contract with Nigeria for patrol craft in 2017.

The SLN also has several in-house research and development centres. For example, the Weapon Engineering Unit was active in adapting weapons stations to meet the navy's specific requirements for countering Sea Tiger attacks during the civil war whilst the Electrical New Design Centre at Trincomalee has been upgrading legacy propulsion and steering control systems on the FACs based there.

FUTURE DIRECTION

In his 2021 New Year's address to naval personnel, the SLN's leader Vice Admiral Ulugetenne remarked that the SLN expected to '... acquire new platforms to the naval fleet this year onwards'. In subsequent remarks he suggested there would be significant investment in larger and more advanced platforms to expand surveillance capabilities. Whilst no further details have been publicly revealed, these observations are very much in line with the SLN's general transition from a brown water navy largely focused on undertaking active combat operations in the littoral towards a constabulary force tasked with ensuring the security of the far reaches of Sri Lanka's EEZ. This journey is likely to require the acquisition of further large OPVs and continued investment in other aspects of maritime domain awareness. It is known that the SLN would like to acquire its own naval air arm but increased collaboration with the SLAF to expand the use of helicopters and UAVs at sea possibly seems more realistic in the short to medium term.

The SLN is unlikely to face any significant threat from state-based actors in the foreseeable future. However, it needs always to be aware of the balancing act that needs to be maintained between neighbouring India and an increasingly omnipresent China as both strive for regional influence. The great-power antagonism between China and the United States also needs to be added to this mix. To date, the SLN has adroitly managed to make friends with all three as it seeks to strengthen its maritime security on a largely collaborative basis, as evidenced by the willingness of these rivals to supply its major warships. However, China has built up enormous economic leverage in Sri Lanka over the years and the SLN may struggle to maintain its policy of non-alignment in the years ahead if local tensions continue to increase.

The SLN's 'advanced offshore patrol vessel' *Sayurala* seen operating with a US Navy MH-60S Seahawk helicopter from the Littoral Combat Ship *Charleston* (LCS-18) in June 2021. Sri Lanka's navy continues to benefit from solid cooperation with other regional navies as it enhances its ability to perform constabulary missions across its EEZ. *(US Navy)*

Having learnt from the mistakes of not having a force effectively configured to meet the asymmetric threat posed by the LTTE's Sea Tigers in the early years of the Eelam Wars, the SLN ultimately proved able to adapt remarkably well to the changing battlespace. This counter-insurgency expertise was, however, won at a high cost in terms of both lives and money. It has also left the SLN with a navy that is arguably neither best equipped nor structured for the very different constabulary mission that is now its primary purpose. Evidence suggests the SLN is making reasonable progress in evolving to equip itself for its new role against a difficult budgetary backdrop. It will continue to rely on evolving maritime domain awareness and its sound cooperation with partner navies as it progresses along this journey.

Notes

1. A much more detailed history of the SLN is provided by the navy's official website at: www.navy.lk/history.html

2. See *Maritime Doctrine of Sri Lanka* (MDSL) – SLN BR 1 (Colombo: Sri Lanka Navy Headquarters, 2020); accessible at: www.navy.lk/assets/images/english/Doctrine/sln_br1_ctp.pdf

3. Some sources differ as to the exact location of the various regional headquarters. There are also considerable variations in the number of subsidiary bases in each Naval Command; for example the South Eastern Command's headquarters at SLNS Mahanaga is the only commissioned SLN base in that Naval Area.

4. Another *Vikram* class OPV, *Suraksha* (formerly *Varuna*), is operated by the SLCG.

5. Formerly known as the 'Arrow' type, the 'Cedric' type patrol craft were renamed in honour of Commander Cedric Martenstyn, one of the founding fathers of Sri Lanka's Special Boat Squadron. Although the design of the boat is commonly attributed to him, some sources suggest captured Sea Tiger craft proved the real inspiration.

6. As mentioned in note 5 above, Commander Cedric Martenstyn is often regarded as one of the SBS's other initial founders.

2.4 REGIONAL REVIEW

EUROPE AND RUSSIA

Author:
Conrad Waters

Naval developments in Europe over the past year have been typified both by continued investment in new naval programmes and ongoing tensions with Russia. Inevitably these two factors have a strong correlation, with concerns over Russia's assertive actions on the world stage encouraging many countries to embark on a programme of rearmament. It is also noteworthy that much equipment acquired towards the end of the Cold War is now falling due for replacement, further driving the need for a renewed period of naval construction.

The last-mentioned factor is certainly having an impact in France, which has announced two large and expensive replacement programmes over the last twelve months. First off the starting blocks was the project for the *porte-avions de nouvelle génération* (PANG). This was formally launched by President Emanuel Macron on a visit to the Framatome reactor components plant at Le Creusot on 8 December 2020. Following on from preliminary design studies that commenced in 2018, the announcement confirmed that the new ship will be powered by two K22 nuclear reactors, a choice partly dictated by a desire to sustain French industrial expertise in this sphere. Other key design characteristics include a displacement of c. 75,000 tons and a length of around 300m, making the ship the largest warship ever built in Europe once she is completed. Although Naval Group will lead the project, actual assembly will take place at the giant Chantiers de l'Atlantique shipyard at Saint Nazaire. Construction is expected to commence in 2025 after completion of detailed design work, allowing the ship to be ready for sea trials around 2036. This should permit a smooth handover from the existing *Charles de Gaulle* before the veteran aircraft carrier retires in 2038.

Subsequently, on 19 February 2021, French defence minister Florence Parly announced the start of the full-scale realisation of the *sous-marin nucléaire lanceur d'engins de troisieme génération* (SNLE 3G) project. This will see four new strategic submarines replace the existing *Le Triomphant* class at five-yearly intervals from around 2035 onwards. Confirmation of the programme is a further major boost for Naval Group – which will build the new boats at its Cherbourg facility – in particular and for the broader French nuclear naval industrial sector in general. However, the vast costs associated with realising these projects in parallel – yet to be officially announced but likely to amount to tens of billions of euros overall – may well ultimately have an adverse impact on the amount of resources available for other French naval modernisation plans.

Meanwhile, an illustration of the new tensions in 'East-West' relations was provided at the end of June 2021 when the British Royal Navy destroyer *Defender* became the focal point of a diplomatic incident after conducting a freedom of navigation operation (FONOP) in disputed waters off the Russian-annexed peninsula of the Crimea. The mission resulted in Russian claims that they had fired warning shots and – more unusually – released bombs to scare off the warship. This interpretation was robustly challenged by the British Ministry of Defence, the absence of any bombing drops being supported by journalists 'fortuitously' embarked on the Royal Navy destroyer. The following day saw further harassment of NATO vessels operating in the Black Sea when the Dutch frigate *Evertsen* – part of the same deployment as the British ship but operating independently at the time – was subject to hours of feint attacks by Russian jets.

There has been much debate over the legitimacy of such NATO operations close to what Russia claims as its own shores. Freedom of navigation has become a tenet of the 'rules based' world order and one that the 'Western' nations seem determined to uphold. Moreover, it seems clear that international law is on HMS *Defender's* side.[1] At the same time, however, it is also evident that such operations are adding heat to an international situation that is already strained, with the potential danger of miscalculation ever present. In this case, both sides seemingly took care to avoid any escalation of the confrontation beyond embarking on a war of words that were likely targeted more at a domestic audience than any enemy. It has to be hoped that similar good sense prevails as the protagonists on both sides of the new Cold War continue to challenge each other's boundaries.

Table 2.4.1 sets out the composition of the major European fleets as at mid-2021.

A computer-generated graphic of the new French *sous-marin nucléaire lanceur d'engins de troisieme génération* (SNLE 3G) strategic submarine. Four of these boats will be delivered from 2035 onwards, inevitably putting a strain on the overall French naval and defence budgets. *(Naval Group)*

TABLE 2.4.1: FLEET STRENGTHS IN WESTERN EUROPE – LARGER NAVIES (MID 2021)

COUNTRY	FRANCE	GERMANY	GREECE	ITALY	NETHERLANDS	SPAIN	TURKEY	UK
Aircraft Carrier (CVN/CV)	1	–	–	1	–	–	–	2
Support/Helicopter Carrier (CVS/CVH)	–	–	–	1	–	–	–	–
Strategic Missile Submarine (SSBN)	4	–	–	–	–	–	–	4
Attack Submarine (SSN)	6	–	–	–	–	–	–	6
Patrol Submarine (SSK)	–	6	11[2]	8	4	2	12	–
Fleet Escort (DDG/FFG)	15	11	13	15	6	11	16	18
Patrol Escort/Corvette (FFG/FSG/FS)	12	5	–	–	–	–	10	–
Missile Armed Attack Craft (PGG/PTG)	–	–	18	–	–	–	19	–
Mine Countermeasures Vessel (MCMV)	13	10[1]	3	10	6	6	11	13
Major Amphibious (LHD/LPD/LPH/LSD)	3	–	–	3	2[3]	3	–	5

Notes:

1 Two further units used as support vessels.

2 Headline figures overstate the actual position, as some old units are of doubtful operational status.

3 Also one joint support ship with amphibious capabilities.

MAJOR REGIONAL POWERS – FRANCE

In addition to the two major long-term projects discussed in the introductory section, France's *Marine Nationale* continues to make headway with a broad spectrum of naval programmes under the framework of the 2019–2025 Military Planning Law. In overall terms, the pandemic has had only minimal impact on the implementation of these projects beyond relatively modest changes to delivery schedules that are, in any event, the frequent lot of modern military procurement. The fact that the modernisation effort is being supported by a relatively generous medium-term defence settlement is also positive, although there remains a residual fear that the cost of high profile and expensive programmes will impact less newsworthy but still vital procurement.

Table 2.4.2 provides further details of fleet composition as of mid-2021 whilst discussion of specific warship categories follows.

Submarines: A highlight of the last year has been delivery of the new attack submarine *Suffren*, first of the 'Barracuda' type, by Naval Group on 6 November 2020. The seemingly trouble-free handover after just six months of sea trials and in the midst of a pandemic speaks highly of the French group's final realisation of a programme that has previously suffered from significant delays. After completing a period of minor defect rectification early in 2020, the boat is expected to continue a programme of operational tests and trials before being formally admitted into naval service before the end of 2021. By that time, *Duguay-Trouin* – second member of the six-boat series – should have been launched from Naval Group's facility at Cherbourg. Subsequently submarines will be completed at roughly biennial intervals, allowing the overall programme to be concluded by the end of the current decade.

The previous delays to the 'Barracuda' project has left the existing fleet of *Rubis* class submarines rather stretched, a situation which has been exacerbated by a destructive fire that broke out aboard *Perle* – the youngest of the class – whilst she was undergoing a final refuelling and modernisation at Toulon in June 2020. In October 2020 it was decided to effect a repair that would involve replacing the boat's fire-damaged forward 'C' and 'D' sections with the equivalent sections from her already decommissioned sister *Saphir*. The process – which involves extending the boat's hull by 1.4m to integrate the junction boxes required to splice the electrical cabling of the constituent parts – is being carried out at Cherbourg under a major engineering and industrial effort that will require around 350,000 hours of additional work. The 'hybridised' boat will then be returned to Toulon to allow the rest of the refuelling and modernisation to be completed.[2]

Major Surface Combatants: The last year has seen further progress with French surface fleet renewal. April 2021 saw delivery of *Alsace*, the first of two FREMM-DA (*Frégate Européenne Multimissions de Défense Aérienne*) variants that form the last units in France's truncated FREMM construction programme. Although not encompassing all the upgrades initially planned for this type, the new ship benefits from modifications that include a uniform armament of four octuple Sylver A50 vertical launch system (VLS) modules, enhancements to the SETIS combat management system and a more powerful 'Herakles' multi-function radar.[3] Her sister, *Lorraine*, was launched in November 2020 and should join the fleet in 2022. Their arrival paves the way for the withdrawal of the final FASM-70/FAA-70 *Georges Leygues* and *Cassard* class frigates. *Jean Bart* of the latter type has already completed her final voyage with the *Marine Nationale* but is reportedly being kept operational in harbour in case required for transfer to Greece as part of France's bid to supply new 'Belh@rra' type frigates.

Work continues on the domestic variant of the 'Belh@rra' – the *Amiral Ronarc'h* class *frégate de défense et d'intervention* at Naval Group Lorient. The lead ship will now be delivered in 2024 – a year later than planned – but subsequent units of the five-ship class are being accelerated to avoid a potential fall in workload at the Brittany shipyard. The new ships will ultimately replace the existing *La Fayette* class, three of which are scheduled to undergo a modernisation programme to keep them fit for service until the end of the decade. *Courbet* is expected to be the first to complete this process at Toulon towards the end of 2021, being followed by *Aconit* and *La Fayette* herself in subsequent years. When coupled

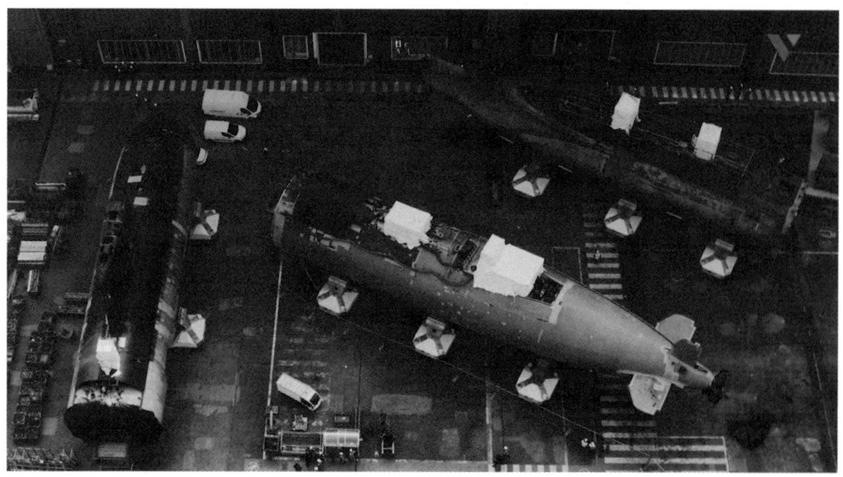

Two views of Naval Group's Cherbourg yard, where the aft end of the French nuclear-powered attack submarine *Perle* has been united with the forward part of her decommissioned sister *Saphir* after her own forward sections were badly damaged by fire during a refit. *Perle*'s damaged forward part can be seen to the left in both images, whilst the integration area for the 'new' submarine is in the centre of the first photograph. *(Naval Group)*

TABLE 2.4.2: FRENCH NAVY: PRINCIPAL UNITS AS AT MID 2021

TYPE	CLASS	NUMBER	TONNAGE	DIMENSIONS	PROPULSION	CREW	DATE
Aircraft Carriers							
Aircraft Carrier – CVN	**CHARLES DE GAULLE**	1	42,000 tons	262m x 33/64m x 9m	Nuclear, 27 knots	1,950	2001
Principal Surface Escorts							
Frigate – FFG	**AQUITAINE** (FREMM)	7[1]	6,000 tons	142m x 20m x 5m	CODLOG, 27 knots	110	2012
Frigate – FFG	**FORBIN** ('Horizon')	2	7,100 tons	153m x 20m x 5m	CODOG, 29+ knots	195	2008
Frigate – FFG	**GEORGES LEYGUES** (FASM-70)	1	4,800 tons	139m x 15m x 5m	CODOG, 30 knots	240	1979
Frigate – FFG	**LA FAYETTE**	5	3,600 tons	125m x 15m x 5m	CODAD, 25 knots	150	1996
Frigate – FSG	**FLORÉAL**	6	3,000 tons	94m x 14m x 4m	CODAD, 20 knots	90	1992
Frigate – FS[2]	**D'ESTIENNE D'ORVES** (A-69)	6	1,300 tons	80m x 10m x 3m	Diesel, 24 knots	90	1976
Submarines							
Submarine – SSBN	**LE TRIOMPHANT**	4	14,400 tons	138m x 13m x 11m	Nuclear, 25 knots	110	1997
Submarine – SSN	**SUFFREN**	1	5,300 tons	100m x 9m x 7m	Nuclear, 25+knots	65	2020
Submarine – SSN	**RUBIS**	5	2,700 tons	74m x 8m x 6m	Nuclear, 25+ knots	70	1983
Major Amphibious Units							
Amph Assault Ship – LHD	**MISTRAL**	3	21,500 tons	199m x 32m x 6m	Diesel-electric, 19 knots	160	2006

Note:
1 Includes one slightly modified *Alsace* variant.
2 Now officially reclassified as offshore patrol vessels.

with the acceleration of the *Amiral Ronarc'h* construction programme, this process holds out the prospect of the French Navy being able to grow its fleet structure beyond the planned force of fifteen 'first rank' frigates in the medium term.

Minor Warships and Auxiliaries: The significant programme of investment required to renew France's front-line force of naval combatants is being driven further upwards by increasingly pressing demands to replace second-line capabilities. Perhaps the most urgent need is for the new *bâtiments ravitailleurs de forces* replenishment tankers, which will be assembled at the Chantiers de l'Atlantique shipyard based on Fincantieri's *Vulcano* design. The use of the Italian ship as a design reference in preference to a Naval Group concept was related to Fincantieri's planned acquisition of the giant Saint Nazaire facility, a decision that France might now regret following the collapse of the deal in January 2021. Sections of the lead tanker, *Jacques Chevallier*, are under construction both in France and at Fincantieri's Castellammare di Stabia facility near Naples, which has been sub-contracted to build sections of the new vessels. However, her delivery has reportedly been put back to 2023, putting further pressure on France's remaining replenishment assets. The force of existing replenishment

The seventh French FREMM frigate *Alsace* was delivered in April 2021. A specialised air-defence variant, she incorporates a number of detailed differences compared with other French members of the class. *(Naval Group)*

oilers has now fallen to just two ships following the retirement of *Var* in mid-2021. This leaves just two *Durance* class vessels in service.

The contract for new *patrouilleurs outre-mer* (POM) is also running slightly behind schedule. First steel cutting for *Auguste Bénébig*, the first of six vessels enlarged from the existing *La Confiance* class, took place at Socarenam Saint Malo in October 2020. However, her delivery is now planned for 2023, a year later than first envisaged. The ships will be followed by ten larger *patrouilleurs océaniques* (PO) that will be built under a programme coordinated by Naval Group but allocated to a number of French shipyards. Reports suggest that the new vessels will be around 90m in length and displace up to 2,000 tons. Although equipped only with a light defensive armament, they will be capable of helicopter operation and be equipped with sonar to perform the current submarine force protection role assigned to the remaining A-69 type *avisos*.[4] This class has now shrunk to six ships following the withdrawal of *Lieutenant de vaisseau Le Hénaff* on 31 July 2020.

A number of projects are somewhat further in the distance. These include replacements for the *Floréal* class patrol frigates under the framework of the European Patrol Corvette (EPC), new hydrographic ships and a future mine countermeasures system – *Système de Lutte Anti-Mines Future* or SLAM-F – that was formally approved in October 2020. The last mentioned will comprise between four and six *bâtiments de guerre des mines* (BGDM) 'mother-ships', eight suites of unmanned drone systems, five *bâtiments bases de plongeurs démineurs de nouvelle génération* (BBPD NG) diving support vessels and a mine warfare data command system.

MAJOR REGIONAL POWERS – ITALY

The Italian Navy's major combatants are summarised in Table 2.4.3. There has been a slight fall in fleet numbers year-on-year following the decommissioning of the *Maestrale* class frigate *Espero* on 30 June 2021. In the normal course of events she would have been replaced by the ninth Italian multi-mission FREMM frigate, the general-purpose variant *Spartaco Schergat*. However, *Schergat* was sold to Egypt along with the tenth unit, *Emilio Bianchi*, in the second half of 2020. Although replacement units have been ordered from Fincantieri, the *Marina Militare* has had to accept a temporary dip in numbers to support the country's export ambitions.

Italy's new replenishment oiler *Vulcano* has now entered service after a protracted build period delayed both by a fire and the COVID-19 pandemic. The Italian Navy hope to build two more vessel of the type for their own fleet and the design also forms the basis for the French Navy's new *Jacques Chevallier* class. *(Fincantieri)*

The main arrival over the last year has been the new replenishment oiler *Vulcano*, which was handed over to the navy on 12 March 2021. Her delivery had been delayed, firstly, by a fire in her bridge structure in July 2018 and then the onset of the COVID-19 pandemic. Her arrival marks a further step in completing the ambitious programme of new construction set out under the 'Naval Law' of 2014/15 following the delivery of two fast interceptors to the navy's Special Forces in 2019–20. Also under construction are the amphibious assault ship *Trieste* – expected to enter service during 2022 to replace the veteran *Giuseppe Garibaldi* – and six out of seven frigate-like PPA (*Pattugliatore Polivalente d'Altura*) multi-purpose offshore patrol ships. Three of these are now in the water following the launch of *Raimondo Montecuccoli* on 13 March 2021. The lead ship, *Paolo Thaon di Revel*, is scheduled to be delivered in September 2021 after completing an extensive series of trials.[5]

With the 'Naval Law' programme now well underway, the Italian Navy is seeking to move forward with the next phase of fleet renewal, which also encompasses significant acquisitions and capability upgrades.[6] The most advanced of these is a plan for four additional Type 212NFS (near future submarines), which essentially combine a slightly stretched ThyssenKrupp Marine Systems (TKMS) Type 212A design with Italian electronics. A €1.35bn (c. US$1.6bn) contract for an initial pair of boats was signed with Fincantieri in February 2021 under the auspices of the European Organisation for Joint Armament Cooperation (OCCAR). Deliveries are scheduled for 2027 and 2029, replacing the third batch of *Sauro* class submarines. A new submarine rescue vessel is to be built by the T Mariotti shipyard in Genoa to support the underwater flotilla, replacing the existing elderly *Anteo*.

Design studies are either underway or imminent for two c. 10,000-ton air-defence destroyers in substitution for the existing pair of *Luigi Durand de la Penne* class ships, as well as for a replacement flotilla of mine countermeasures vessels (MCMVs). In February 2021, Intermarine was awarded a project definition study for the latter programme

Table 2.4.3: ITALIAN NAVY: PRINCIPAL UNITS AS AT MID 2021

TYPE	CLASS	NUMBER	TONNAGE	DIMENSIONS	PROPULSION	CREW	DATE
Aircraft Carriers							
Aircraft Carrier – CV	**CAVOUR**	1	27,100 tons	244m x 30/39m x 9m	COGAG, 29 knots	800	2008
Aircraft Carrier – CVS	**GIUSEPPE GARIBALDI**[1]	1	13,900 tons	180m x 23/31m x 7m	COGAG, 30 knots	825	1985
Principal Surface Escorts							
Frigate – FFG	**CARLO BERGAMINI** (FREMM)[2]	8	6,700 tons	144m x 20m x 5m	CODLOG, 27 knots	145	2013
Frigate – FFG	**ANDREA DORIA** ('Horizon')	2	7,100 tons	153m x 20m x 5m	CODOG, 29+ knots	190	2007
Destroyer – DDG	**DE LA PENNE**	2	5,400 tons	148m x 16m x 5m	CODOG, 31 knots	375	1993
Frigate – FFG	**MAESTRALE**	3	3,100 tons	123m x 13m x 4m	CODOG, 30+ knots	225	1982
Submarines							
Submarine – SSK	**TODARO** (Type 212A)	4	1,800 tons	56m x 7m x 6m	AIP, 20+ knots	30	2006
Submarine – SSK	**PELOSI** (Improved SAURO)	4	1,700 tons	64m x 7m x 6m	Diesel-electric, 20 knots	50	1988
Major Amphibious Units							
Landing Platform Dock – LPD	**SAN GIORGIO**	3	8,000 tons	133m x 21m x 5m	Diesel, 20 knots	165	1987

Note:
1 Now operates largely as a LPH.
2 Class includes *Bergamini* (GP) and *Fasan* (ASW) variants.

that could ultimately comprise four *Cacciamine Nuova Generazione-Altura* ocean-going minehunters and eight *Cacciamine Nuova Generazione-Costieri* coastal vessels. Other procurement plans encompass new hydrographic ships, additional *Vulcano* class replenishment oilers and a new series of OPVs. The latter will be acquired under the EU's European Patrol Corvette (EPC) initiative in conjunction with France, Spain and – perhaps – Greece, replacing plans for further PPAs.

MAJOR REGIONAL POWERS – SPAIN

As illustrated by Table 2.4.4, there has been no significant change to the main components of the Spanish Navy over the last year. The auxiliary fleet has, however, been bolstered by acquisition of the roll-on/roll-off cargo vessel *Galicia*, which was renamed *Ysabel* after her delivery in June 2021. The new logistic support ship replaces two smaller vessels – *El Camino Español* and *Martín Posadillo* – that were withdrawn from service in 2019 and 2020. She will be used to provide transport services to the garrisons in Spain's offshore islands and African enclaves.

The main event of the past year has been the long-awaited launch of the lead S-80 Plus submarine *Isaac Peral*. An official launching ceremony was conducted in the presence of King Felipe VI at Cartagena on 22 April 2021 and the submarine was subsequently floated out in a complex operation

The lead S-80 Plus submarine *Isaac Peral* pictured shortly before being rolled out of her building hall for naming and subsequent launch in the first half of 2021. *(Navantia)*

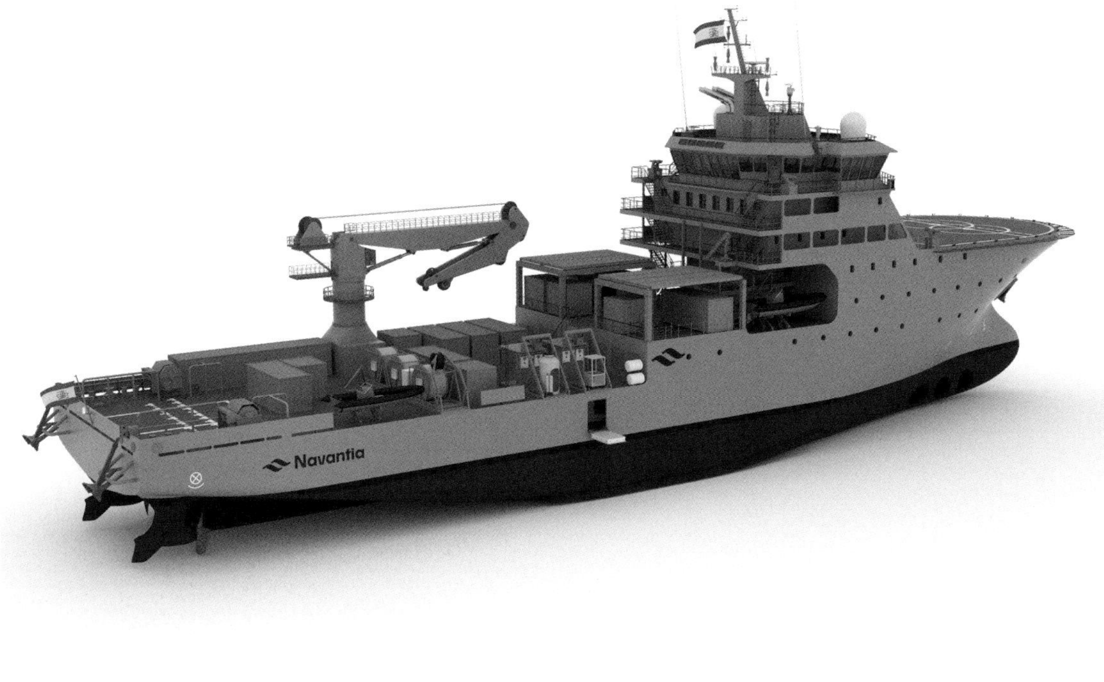

A Navantia graphic of the new BAM-IS submarine rescue vessel, which has recently been authorised by the Spanish government. The ship is derived from the BAM offshore patrol vessels that are already in service with the Spanish Navy. *(Navantia)*

early in May. She is expected to commence sea trials in the course of 2022 and be delivered the following year. Three sister boats remain at various stages of construction at Cartagena. Other positive and related news has been the official approval for a BAM-IS submarine rescue ship that will support the new underwater flotilla. The new vessel is expected to be officially ordered in the second half of 2021 and will ultimately replace the veteran *Neptuno.*

A detailed review of the *Armada Española* is contained in Chapter 2.4A.

MAJOR REGIONAL POWERS – UNITED KINGDOM

Current Royal Navy strength is set out in Table 2.4.5. Although the last twelve months have provided grounds for the Royal Navy to be optimistic over its longer-term development, fleet strength has actually fallen in the short term. The *Trafalgar* class attack submarine *Trenchant* entered the naval base at Devonport for the last time in March 2021, at least a year before *Anson*, her replacement, is likely to be delivered. In addition, decisions taken as a result of the Integrated Review mean that the Type 23 frigate *Monmouth* – laid up pending modernisation since 2019 – will never again see service under the White Ensign. The review also marked the end of the line for the auxiliaries *Fort Austin* and *Fort Rosalie*. Already laid up in reserve, they were both officially withdrawn from service on 31 March 2021.

More positive developments have included completion of deliveries of the 'River' Batch 2 OPVs and the return to sea of the aircraft carrier *Prince of Wales* after an extended period alongside due to accidental flooding that damaged her electrical installation. Work on the third Type 26 frigate, *Belfast*, commenced at the end of June 2021. At the other end of the size scale two 19m fast patrol boats – named *Cutlass* and *Dagger* – were ordered from Liverpool-based Marine Specialised Technology in July 2020 to serve with the Gibraltar Squadron. Both should be delivered before the first quarter of 2022. A positive development of a different kind was the selection of Commodore Jude Terry for promotion to Rear Admiral, the first time in the centuries-long history of the Royal Navy that a woman has held the post. The new admiral is scheduled to assume her role as the Royal Navy's Director of People and Training and Naval Secretary early in 2022.

A much more detailed review of the Royal Navy's future direction is provided in Chapter 2.4B.

MID-SIZED REGIONAL FLEETS

Germany: The last year has seen Germany's *Deutsche Marine* continue to benefit from existing procurement projects whilst launching new programmes. Prominent amongst the former is the contract for four F125 *Baden-Württemberg* class stabilisation frigates. The third of these, *Sachsen-Anhalt*, was commissioned at Wilhelmshaven on 17

Table 2.4.4: SPANISH NAVY: PRINCIPAL UNITS AS AT MID 2021

TYPE	CLASS	NUMBER	TONNAGE	DIMENSIONS	PROPULSION	CREW	DATE
Principal Surface Escorts							
Frigate – FFG	**ÁLVARO DE BAZÁN** (F-100)	5	6,300 tons	147m x 19m x 5m	CODOG, 28 knots	200	2002
Frigate – FFG	**SANTA MARÍA** (FFG-7)	6	4,100 tons	138m x 14m x 5m	COGAG, 30 knots	225	1986
Submarines							
Submarine – SSK	**GALERNA** (S-70/AGOSTA)	2	1,800 tons	68m x 7m x 6m	Diesel-electric, 21 knots	60	1983
Major Amphibious Units							
Amph Assault Ship – LHD	**JUAN CARLOS I**	1	27,100 tons	231m x 32m x 7m	IEP, 21 knots	245	2010
Landing Platform Dock – LPD	**GALICIA**	2	13,000 tons	160m x 25m x 6m	Diesel, 20 knots	185	1998

Table 2.4.5: BRITISH ROYAL NAVY: PRINCIPAL UNITS AS AT MID 2021

TYPE	CLASS	NUMBER	TONNAGE	DIMENSIONS	PROPULSION	CREW	DATE
Aircraft Carriers							
Aircraft Carrier – CV	**QUEEN ELIZABETH**	2	65,000 tons	284m x 73m x 11m	IEP, 26 knots+	1,600	2017
Principal Surface Escorts							
Destroyer – DDG	**DARING** (Type 45)	6	7,500 tons	152m x 21m x 5m	IEP, 30 knots	190	2008
Frigate – FFG	**NORFOLK** (Type 23)	12	4,900 tons	133m x 16m x 5m	CODLAG, 30 knots	185	1990
Submarines							
Submarine – SSBN	**VANGUARD**	4	16,000 tons	150m x 13m x 12m	Nuclear, 25+ knots	135	1993
Submarine – SSN	**ASTUTE**	4	7,800 tons	93m x 11m x 10m	Nuclear, 30+ knots	100	2010
Submarine – SSN	**TRAFALGAR**	2	5,200 tons	85m x 10m x 10m	Nuclear, 30+ knots	130	1983
Major Amphibious Units							
Landing Platform Dock – LPD	**ALBION**	2[1]	18,500 tons	176m x 29m x 7m	IEP, 18 knots	325	2003
Landing Ship Dock – LSD (A)	**LARGS BAY**	3	16,200 tons	176m x 26m x 6m	Diesel-electric, 18 knots	60	2006

Notes:
1. One at extended readiness.

The Royal Navy *Trafalgar* class submarine *Trenchant* returning to Devonport to pay off for the final time in March 2021. Delays to the replacement *Astute* class programme mean that it will be some time before her replacement *Anson* – launched in April 2021 – is operational. *(Crown Copyright 2021)*

Sachsen-Anhalt, Germany's third F125 class frigate, silhouetted in the morning dawn against the backdrop of the Laboe Naval Memorial near Kiel. She was commissioned on 17 May 2021. *(TKMS)*

As has been the case for many of the German Navy's recent surface combatants, the second batch of K130 corvettes are being built in sections at various shipyards. Integration of this batch is being focused on the Blohm & Voss yard at Hamburg. *(ARGE K130 via Lürssen)*

May 2021. She is likely to be followed into service by the final unit, *Rheinland-Pfalz*, by the end of the year. The follow on MKS-180 multi-role combat ship – now officially designated the F126 class frigate – remains in the detailed design stage prior to planned commencement of fabrication in 2023. The four new ships – with a displacement of nearly 10,000 tons, a length of 166m and a beam of 21m – will be the largest surface combatants in Western Europe and Germany's biggest since the Second World War. Although contracted to Damen, actual construction will take place in Germany at Lürssen's yards, which should deliver the first of the new vessels by 2028. Shipyard utilisation between the two classes is being bridged by the 2017 contract for a second batch of five K130 *Braunschweig* class corvettes. These are being constructed in two halves at various German yards before being shipped to Hamburg for integration at Lürssen's Blohm & Voss facility.

In June 2021, the German parliament maintained the recent momentum achieved in the process of naval renewal by providing budgetary approval for several new programmes. By far the most important of these is the project for Type 212CD submarines, which will be jointly procured by Germany and Norway as a variant of the existing Type 212A design. The detailed contractual and technical provisions for the programme were previously agreed with builders TKMS in March 2021, with construction work scheduled to begin in 2023. The German Navy is to benefit from two new submarines under the six-boat programme, with deliveries scheduled for 2032 and 2034.

The new budgetary approvals also encompass a number of auxiliary ships, including two new c. 20,000-ton Type 707 fleet replenishment oilers to replace the existing Type 704 class vessels *Röhn* and *Spessart* and three Type 424 intelligence-gathering vessels. The latter were contracted with Lürssen on the day the budgetary approvals were announced and will renew the current flotilla of 1980s-era Type 423 *Oste* on a like-for-like basis from 2027. Also moving forward is the purchase of five Poseidon P-8A maritime patrol aircraft, making Germany the third European country after the United Kingdom and Norway to adopt the aircraft.

Greece: A decade or more of financial constraints have had a severe impact on the Hellenic Navy. Although the fleet has remained broadly constant in

The Hellenic Navy took delivery of its long-awaited sixth *Roussen* class fast attack craft, *Ypoploiarchos Karathanasis*, in July 2020 and saw the launch of its seventh, *Ypoploiarchos Vlachakos*, the following month. *(Hellenic Navy)*

numerical terms, overall effectiveness has dropped considerably as major warships have aged without modernisation. Deliveries of the quartet of Type 214HN AIP-equipped submarines between 2010 and 2016 has left the underwater flotilla with a core of relatively modern boats. However, the four MEKO 200HN *Hydra* and nine former Dutch *Elli* ('Standard') class frigates that form the core of the surface fleet are all well in excess of 20 years old and in urgent need of modernisation or replacement.

Increased tensions with Turkey over maritime resources are now spurring on the Greek government to remedy the worst deficiencies.[7] In September 2020 Greek Premier Kyriakos Mitsotakis announced a major investment programme for the country's armed forces. These included combining previously-announced plans for upgrades to the *Hydra* class with acquisition of a new class of multi-role surface combatant. In late 2019, Greece had signed a letter of intent to purchase two 'Belh@rra' export variants of Naval Group's *frégate de défense et d'intervention* (FDI). However, the new contract has been thrown open to a wider range of potential partners. As of mid-2021, bids from France (FDI), Germany (MEKO A-200), Italy (FREMM), the Netherlands ('Sigma 11515'), the United Kingdom ('Arrowhead 140') and the United States (LCS-1 *Freedom* derived multi-mission combatant) reportedly remained under consideration for a contract that should be awarded before the end of 2021. A key factor in the eventual decision will be the industrial strategy adopted to support local construction given the priority the government likely attaches to revitalising the domestic Hellenic and Elefsis shipyards under new ownership.

A number of more modest enhancements to the fleet are currently underway. In October 2020, a contract for four MH-60R Seahawk helicopters was placed with Lockheed Martin. The following month saw a long-delayed agreement to acquire thirty-six SeaHake mod 4 torpedoes from Atlas Elektronik to equip the Type 214 submarines. Long delayed completion of the remaining *Roussen* fast attack craft is also finally moving forward. The sixth member of the class, *Ypoploiarchos Karathanasis*, was finally commissioned on 28 July 2020 whilst the seventh, *Ypoploiarchos Vlachakos*, was launched at the Elefsis yard the following month. There have been reports in the Greek press that at least one more of the class will be built. However, a proposal from Israel Shipyards for the licensed production of a 'Themistocles' variant of its Sa'ar 72 design provides another potential option to maintain production at a strategically important facility.

The only major departure from the fleet has been the former British Royal Navy 'Hunt' class MCMV *Kallisto* (formerly *Berkley*). The minehunter was in collision with the container vessel *Maersk Launceston* off the Port of Piraeus on 27 October 2020, losing most of her stern aft of the funnel. Although the forward half of the ship was towed into port, it seems unlikely she will be repairable. Fortunately there were no fatalities in the collision.

The Netherlands and Belgium: The Belgian Naval Component and the Royal Netherlands Navy (RNLN) have made further progress with joint acquisition programmes over the last year. The project for twelve new mine countermeasures 'motherships' – six for each navy – being taken forward under the so-called BE-NL rMCM Replacement Mine Countermeasures Capability Programme is most advanced. The pandemic is only having a modest impact on design milestones agreed with the Belgium Naval Robotics Consortium tasked with implementing the project and work on the first vessel is expected to commence in mid-2021. Construction is being carried out by the Kership joint venture between Naval Group and Piriou, which expects to deliver the lead unit – destined for Belgium – before the end of 2024. Procurement of the 'toolbox' of largely manned and autonomous systems that will be deployed from the c. 2,800-ton vessels is also well in hand. A start has already been made to withdraw existing vessels, with the veteran 1966-delivered command and support ship *Godetia* being decommissioned from Belgian service on 26 June 2021.[8]

Deliveries of ships under the closely-linked navies' other joint project – for two pairs of 'ASWF' anti-submarine warfare optimised frigates – are further in the future. A construction contract for the vessels is expected to be awarded to Damen by the end of

2021. The two Dutch frigates should be ready for operational service with the navy in 2028 and 2029 after completion of post-delivery trials. They will be followed by the Belgian ships in 2030. Reports suggest that the latter pair are being subject to 'optimisation' to keep the programme within budget, a process that has included halving their vertical launch capacity. Although both Belgian and Netherlands defence budgets are increasing, funding still remains tight. Notably, the last few months have seen the RNLN undertake cost-saving measures such as deferring maintenance and temporarily laying up ships, the latter step also apparently being driven by personnel shortages. To some extent, it seems that funding being made available for new investment is not being replicated in the money provided for the operational fleet.

National Dutch shipbuilding programmes are also making headway. Fabrication of the new combat support ship *Den Helder* commenced at Damen's facility at Galati in Romania in December 2020. This was followed by a formal keel-laying ceremony on 2 June 2021. Completion remains scheduled for 2024. The much more significant project for four new submarines to replace the existing *Walrus* class is still officially scheduled to see deliveries commence in 2028. Three rival groups headed by Naval Group, Saab Kockums and TKMS are bidding to build the new boats but a decision as to which to select has now been postponed until the end of 2022. This makes achieving the existing deadline seem increasingly unlikely.[9]

Turkey: Arguably the leading navy in the Eastern Mediterranean, Turkey continues to maintain a powerful fleet of both seagoing and coastal surface combatants supplemented by Europe's largest submarine flotilla. The Turkish Navy's development has been supported by an increasingly proficient national naval shipbuilding and equipment industry. This is demonstrating an ability to undertake its own construction programmes with only a diminishing amount of external assistance. An increased indigenous capability is proving particularly valuable against a backdrop of more challenging conditions for international collaboration given maritime boundary disputes with neighbours in the Eastern Mediterranean and strains with the United States over the country's acquisition of Russian air-defence systems.

The most noteworthy development over the past year has been the launch of the frigate *Istanbul* at Istanbul Naval Shipyard on 23 January 2021. She is the first of four planned 'Istif' class ships. A stretched variant of the previous 'Ada' class corvette 'MILGEM' national surface ship programme, *Istanbul* will incorporate a greater proportion of national equipment. Prominent amongst this will be locally-designed MDAS *Milli Dikey Atis Sistemi* VLS. Originally planned to be introduced in the planned TF-2000 air-warfare destroyers, its production has reportedly been brought forward to compensate for an American ban on import of the Mk 41 VLS.[10] It is not altogether clear which weapon will be housed in the launch cells, although a variant of the indigenous 'Hisar' surface-to-air missile seems possible. In any event, *Istanbul* will be equipped with the new nationally-produced 'Atmaca' surface-to-surface missile, which is a replacement for the existing American Harpoon. The new missile completed a final test firing against the retired salvage and rescue vessel *Işın* on 18 June 2021, clearing the way for serial production.

There has been little news of developments with respect to the TF-2000 programme over the past year. This may reflect a desire to prioritise the current modernisation of the existing quartet of MEKO 200TN Batch 2 *Barbaros* class frigates by an Aselsan-Havelsan consortium. The design phase of this upgrade was completed in 2020 and it is hoped to return the first modernised ship to service by the end of 2022. The upgrades encompass the substitution of indigenous systems for much 'legacy' foreign technology. Notably, the existing Thales combat management system (CMS) will be replaced by B-SYS, the latest iteration of the national GENESIS CMS. The aim, therefore, may well be to prove the

The new Turkish amphibious assault ship *Anadolu* pictured in the course of being fitted out amongst the cranes of the Sedef Shipbuilding yard in Istanbul in March 2021. The Turkish government are desperately trying to find ways of deploying fixed-wing aircraft from the carrier after the blocking of F-35B strike fighter purchases by the United States. *(Devrim Yaylali)*

new locally-built equipment on the modernised frigates before progressing to building the sophisticated and complex destroyers.

Another important indigenous project currently underway is the FAC55 fast attack craft programme. These ships will be constructed on the basis of a concept being developed by STM – the state-owned defence and security conglomerate also responsible for the 'MILGEM' design – under a 24-month contract agreed in August 2020. As the Turkish Navy's existing FACs are all of German Lürssen design, this marks another step forward in expanding local capabilities.

Some Turkish warship construction still depends on external support. This includes the programme for Type 214TN AIP-equipped submarines, which relies heavily on TKMS technology. Six of these boats are being built at the Gölcük Naval Shipyard under a contract first agreed back in 2009 but implementation has been slow, possibly due the incorporation of significant amounts of Turkish-manufactured equipment. The first unit, *Piri Reis*, was floated out in March 2021 after a formal 'launch' ceremony held in December 2019 and should be in service by 2022. Subsequent members of the class are expected to follow at annual intervals and could then be followed by a fully national 'MILDEN' submarine. Meanwhile, construction of the new amphibious assault ship *Anadolu* – based on Spain's *Juan Carlos I* design – is reportedly close to completion. The Turkish Navy previously had hopes of operating the ship in the light aircraft carrier role but the United States' cancellation of orders for F-35B strike fighters has put paid to this ambition, at least temporarily. Reports have suggested the navy is looking to adapt Bayraktar series UAVs or even the developmental Hürjet jet trainer to operate from the new ship in a sign of just how important membership of the exclusive 'carrier club' is to President Erdoğan's nationalist administration.[11]

Romania's Type 22 frigates are due to be modernised by Naval Group but it is unclear whether or not a firm contract has been signed. This photograph shows *Regele Ferdinand* – the former HMS *Coventry* – during an exercise in the Black Sea in July 2020. *(Crown Copyright 2020)*

OTHER REGIONAL FLEETS

Black Sea and Mediterranean: Tensions with Russia are fuelling a surge of procurement activity in the Black Sea. The last year has seen particularly significant developments with respect to **Ukraine**, which is attempting to rebuild its navy following the loss of both ships and supporting infrastructure as a result of Russia's seizure of the Crimea. The most far-reaching but, as yet, unfinalised plan is a draft agreement with the United Kingdom that encompasses the joint construction of *Roussen* type fast attack craft, the transfer of surplus *Sandown* class minehunters and substantial upgrades to base infrastructure. Other planned programmes include shared construction of two 'Ada' class corvettes with Turkey's Istanbul Naval Shipyard and acquisition of United States-built Mk VI patrol craft from SAFE Boats International. In January 2021, it was confirmed that long-lead items for the first pair of what could ultimately be sixteen craft had been ordered, with deliveries scheduled for the end of 2022.

Bulgaria's long-delayed plans for new surface combatants have also been moving forward. In November 2020, a c. US$600m contract was signed with Germany's Lürssen for two multi-role corvettes enlarged from the OPV-80 design already supplied to Australia and Brunei. The ships will be assembled by the local MTG Dolphin shipyard and incorporate a Saab combat management system. Deliveries are expected by the middle of the decade. In the meantime, the MCMV force has already been bolstered by the acquisition of the laid-up Dutch 'Tripartite' class vessels *Hellevoetsluis* (now *Struma*) and *Maassluis* (*Mesta*), which join the former Belgian *Myosotis* (*Tsibar*) to create a modern mine countermeasures flotilla. There has, however, been less news from **Romania,** where the July 2019 announcement of a major deal with Naval Group for four new 'Gowind' corvettes and the modernisation of the country's two former British Royal Navy Type 22 frigates does not yet appear to have resulted in a firm contract.

Away from the Black Sea, **Croatia's** order for four 200-ton *Omiš* class patrol boats following the success of the prototype delivered in 2018 saw a first steel cutting ceremony held at the Brodosplit shipyard in Split responsible for their completion in August 2020. Deliveries are scheduled for between 2022 and 2023. Elsewhere in the Mediterranean, the expansion of **Malta's** small patrol flotilla took a major step forward with the launch of the new 1,800-ton OPV that has been ordered from the Italian Cantiere Navale Vittoria shipyard on 27 February 2021.

The Atlantic and Northern Europe: Looking firstly at the Atlantic, **Portugal's** modernisation of its two 'M' class frigates in the Netherlands is proving more protracted than initially expected. As of mid-2021,

The 31-year-old Norwegian *Ula* class submarine *Uredd* seen during exercises in June 2021. Replacement in the form of the new German Type 212CD design is on the way. *(Norwegian Armed Forces)*

The frigate *Helge Ingstad* pictured making her final voyage to the shipbreakers in February 2021 after it was determined the damage sustained in her 2018 collision and sinking was beyond economic repair. The report into her loss makes hard reading for the Royal Norwegian Navy. *(Maria Gran/Norwegian Armed Forces)*

Bartolomeu Dias (previously *Van Nes*) – the first ship to pass through the modernisation programme – had yet to complete the work package at Den Helder after delays caused an outbreak of COVID-19 onboard the ship and minor damage sustained in collision with a Dutch tug. With her sister *Dom Francisco de Almeida* (the former *Van Galen*) also at Den Helder undergoing the same series of upgrades and the lead MEKO 200PN *Vasco da Gama* laid up in Portugal awaiting her own modernisation, the Portuguese Navy's operational frigate flotilla is down to just two ships. More positively, May 2021 saw formal approval for the acquisition of a third batch of six *Viana do Castelo* offshore patrol ships, taking the class to ten units. The €352m (c. US$425m) project – previously outlined in the 2019-2030 Military Programming Law – will help ensure continuity of employment at the West Sea Viana shipyard that built the early members of the class. The latest ships will include enhanced modularity to give them greater frontline capabilities in areas such as submarine detection and mine countermeasures.

Ireland also has plans for further naval acquisitions, most notably a c. €200m project for a multirole vessel to replace the existing flagship *Eithne* that was first flagged in Ireland's 2015 Defence White Paper. The Irish Naval Service has also reportedly been exploring the purchase of New Zealand's decommissioned inshore patrol vessels *Rotoiti* and *Pukaki* to strengthen post-Brexit fishing patrols in the Irish Sea. The New Zealand ships have lower crewing requirements than the two *Peacock* class patrol vessels currently available for coastal service, an important consideration given the continued shortage of personnel that is afflicting naval operations.[12]

Moving north to Scandinavia, **Norway's** most important programme is for the acquisition of four Type 212CD submarines under the joint project with Germany already mentioned above. An agreement on the construction plan was agreed with TKMS in March 2021 and the receipt of German parliamentary approval for the acquisition in June should allow contracts to be awarded imminently. Norway has an increasingly pressing need to replace its c. 30-year-old *Ula* class boats and will start receiving its units in 2029, three years before the start of deliveries to Germany. Meanwhile, the Norwegian Safety Investigation Authority's part two report into the loss of the frigate *Helge Ingstad* after its collision with the tanker *Solas TS* in the

Hjeltefjord on 8 November 2018 determined that lack of training and preparedness had been significant contributory factors to the ship's failure to survive the accident. Although the collision resulted in damage over and above that which the ship was designed to withstand – affecting five of thirteen watertight compartments – the report concluded that the ship could have been saved if effective steps to maintain stability had been taken.[13]

In neighbouring **Sweden,** publication of the *Totalförsvaret* (Total Defence) 2021–2025 bill in December 2020 confirmed plans for significant naval expansion as part of a broader uplift in military spending in response to the more challenging security situation in the Baltic.[14] The existing five *Visby* class corvettes will receive a major midlife modernisation to maintain them in service until 2040. Additionally, the pair of *Gävle* class corvettes and two *Stockholm* class patrol vessels will be subject to phased replacement over the coming decade, with a project definition phase for two '*Visby* Generation 2' class vessels to replace the former launched in January 2021. Another significant decision confirms plans for a five-strong submarine flotilla through modernisation of the third *Gotland* class submarine, *Halland.* Their sister *Uppland* has now completed the upgrade programme, re-joining *Gotland* in the operational fleet in December 2020.

Finland's *Merivoimat* is progressing its own 'Squadron 2020' upgrade, with a number of long-lead items for the four new *Pohjanmaa* class corvettes being acquired under this programme ordered in the last year. Actual construction is expected to begin in 2022. Modernisation of the existing four *Hamina* class fast attack craft – including the installation of improved anti-submarine capabilities – that is also associated with 'Squadron 2020' continues to make good headway. *Hamina* herself became the second vessel to complete the upgrade in December 2020 and the remaining ships should be completed by the end of 2021. At the other end of the Baltic, **Denmark** has reclassified its two *Absalon* class support vessels as frigates in advance of implementation of a programme to upgrade their anti-submarine warfare capabilities. In a longer-term move, the Danish Ministry of Defence has also launched preliminary studies into a new class of multi-role patrol vessels that are anticipated to incorporate a significant quantity of green technology.

Poland's stop/start naval modernisation has made little tangible progress in the past year. The two remaining former Norwegian Type 207 Kobben class submarines are being withdrawn from service to leave the 'Kilo' class *Orzeł* as the country's sole underwater asset. However, negotiations with Sweden to lease a modified A17 *Södermanland* class submarine as an interim measure pending new procurement have yet to reach a satisfactory conclusion on cost grounds. The current priority appears to be acquisition of three frigate-like 'Miecznik' (Swordfish) coast defence vessels, which will probably be assembled in a Polish shipyard to a foreign design. Pending commencement of the new project, the most important ongoing programme is for three 'Kormoran II' MCMVs. Second of class, *Albatros*, is now undergoing sea trials whilst the third, *Mewa*, was launched on 17 December 2020. Elsewhere in the region, the Baltic Republics also continue to focus on their mine warfare forces, with a €20m upgrade of three of Latvia's 'Tripartite' class minehunters by ECA Group the latest in a series of upgrades undertaken by the three navies.

RUSSIA

At the start of July 2021, Russian President Vladimir Putin endorsed a new National Security Strategy, the first since 2015. The document was notable in being a mirror-image of many of its recent 'Western' counterparts, identifying the threat posed by these unfriendly countries to isolate and destabilise the Russian Federation to protect their own hegemony. The strategy indicated that Moscow considers it legitimate to take both symmetric and asymmetric measures to thwart its opponents, a statement reflective of a similar approach in Russian military doctrine. This can clearly be seen in the recent development of Russian naval forces, where the legacy Soviet fleet of large surface combatants intended to go toe-to-toe in engaging its NATO adversaries is increasingly being replaced by more nimble, diverse assets optimised to exploit its opponents' vulnerabilities.

An important part of this approach is the development of a new generation of powerful strike weapons. The 3M-54 Kalibr (NATO: SS-N-27

Two views of the first series-produced Project 855M 'Yasen-M' class nuclear-powered attack submarine *Kazan*, taken shortly after she was commissioned on 7 May 2021. Missile-armed submarines play a large part amongst the range of nimble and diverse assets optimised to exploit an enemy's vulnerabilities that underpin Russia's National Security Strategy. *(Russian Ministry of Defence)*

'Sizzler'/SS-N-30) series of cruise missiles has already demonstrated the Russian Navy's ability to perform long-range attacks from a dispersed fleet of small and relatively cheap platforms during the course of the Syrian Civil War. New, more potent weapons under development include the autonomous 2M-39 Poseidon intercontinental nuclear-powered and nuclear-armed torpedo and the 3M-22 Tsirkon (NATO: SS-N-33) hypersonic cruise missile. The latter has spent much of the last year undergoing a series of trials from the lead Project 22350 *Admiral Gorshkov* class frigate following an initial test launch from the ship in January 2020. The reportedly c. Mach 8, 500-nautical-mile ranged missile is expected to be widely deployed once it is cleared for operational service, posing a significant headache for Russia's foes.

A summary of Russia's main combatants is provided in Table 2.4.6. As in previous years, this only tells part of the story given the increasing importance attached to the fleet's smaller warships. Some additional information on recent acquisitions of these vessels is contained in the discussion of specific warship types that follows.

Submarines: The Russian Navy's underwater flotilla – particularly its strategic submarine force – remains its top priority. The last year has seen the launch of *Knyaz Oleg*, the fifth member of the 'Borey' class and second Project 995A variant from the Sevmash facility at Severodvinsk. The boat was undergoing sea trials as of mid-2021 and should become operational within the next year. Sevmash is currently building a further three Project 955As and reportedly holds contracts for two further vessels. The last year has also seen the Sevmash yard deliver the lead Project 855M 'Yasen-M' class nuclear-powered attack submarine *Kazan*, which was commissioned on 7 May 2021. An improved version of the prototype Project 855 'Yasen' *Severodvinsk*, she had previously undergone the best part of three years of sea trials and is expected to undertake further tests, this time with the Tsirkon missile, later in 2021. Reportedly far quieter than previous generations of Russian submarines, a further seven Project 855M boats are at various stages of construction at Sevmash following keel-laying ceremonies for *Voronezh* and *Vladivostok* in July 2020. Their timely arrival is crucial to the maintenance of Russia's underwater capabilities given the increasing age of the fleet's remaining Soviet-era boats. Although Project 949A 'Antey' ('Oscar II') and Project 971 'Shchuka-B' ('Akula') continue to be cycled through midlife upgrades, the older surviving members of these classes have now passed their thirtieth birthdays and cannot be expected to remain operational indefinitely.

TABLE 2.4.6: RUSSIAN NAVY: SELECTED PRINCIPAL UNITS AS AT MID 2021

TYPE	CLASS	NUMBER[1]	TONNAGE	DIMENSIONS	PROPULSION	CREW	DATE
Aircraft carriers							
Aircraft Carrier – CV	Project 1143.5 **KUZNETSOV**	1	60,000 tons	306m x 35/73m x 10m	Steam, 32 knots	2,600	1991
Principal Surface Escorts							
Battlecruiser – BCGN	Project 1144.2 **KIROV**	2	25,000 tons	252m x 29m x 9m	CONAS, 32 knots	740	1980
Cruiser – CG	Project 1164 **MOSKVA** ('Slava')	3	12,500 tons	186m x 21m x 8m	COGAG, 32 knots	530	1982
Destroyer – DDG	Project 956/956A **SOVREMENNY**	c. 3	8,000 tons	156m x 17m x 6m	Steam, 32 knots	300	1980
Destroyer – DDG	Project 1155.1 **CHABANENKO** ('Udaloy II')	1	9,000 tons	163m x 19m x 6m	COGAG, 29 knots	250	1999
Destroyer – DDG	Project 1155 **UDALOY**	c. 7	8.500 tons	163m x 19m x 6m	COGAG, 30 knots	300	1980
Frigate – FFG	Project 22350 **GORSHKOV**	2	5,500 tons	135m x 16m x 5m	CODAG, 30 knots	210	2018
Frigate – FFG	Project 11366M **GRIGOROVICH**	3	4,000 tons	125m x 15m x 4m	COGAG, 30 knots	200	2016
Frigate – FFG	Project 1154 **NEUSTRASHIMY**	2	4,400 tons	139m x 16m x 6m	COGAG, 30 knots	210	1993
Frigate – FFG	Project 1135 **BDITELNNY** ('Krivak I/II')	c. 2	3,700 tons	123m x 14m x 5m	COGAG, 32 knots	180	1970
Frigate – FFG	Project 20385 **GREMYASHCHIY**	1	2,500 tons	106m x 11m x 5m	CODAD, 27 knots	100	2020
Frigate – FFG	Project 20380 **STERGUSHCHIY**	7	2,200 tons	105m x 11m x 4m	CODAD, 27 knots	100	2008
Frigate – FFG	Project 11611 **TATARSTAN** ('Gepard')	2	2,000 tons	102m x 13m x 4m	CODOG, 27 knots	100	2002
Submarines							
Submarine – SSBN	Project 95/955A **YURY DOLGORUKY** ('Borey')	4	20,000+ tons	170m x 13m x 10m	Nuclear, 25+ knots	110	2010
Submarine – SSBN	Project 941 **DONSKOY** ('Typhoon')	1	33,000 tons	173m x 23m x 12m	Nuclear, 26 knots	150	1981
Submarine – SSBN	Project 677BDRM **VERKHOTURYE** ('Delta IV')	6	18,000 tons	167m x 12m x 9m	Nuclear, 24 knots	130	1985
Submarine – SSBN	Project 677BDR **ZVEZDA** ('Delta III')	1	12,000 tons	160m x 12m x 9m	Nuclear, 24 knots	130	1976
Submarine – SSGN	Project 855 **SEVERODVINSK** ('Yasen')	2	13,500 tons+	120m x 14m x 9m	Nuclear, 30+ knots	90	2013
Submarine – SSGN	Project 949A ('Oscar II')	c. 7	17,500 tons	154m x 8m x 9m	Nuclear, 30+ knots	100	1986
Submarine – SSN	Project 971 ('Akula I/II')	c. 10	9,500 tons	110m x 14m x 10m	Nuclear, 30+ knots	60	1986
Submarine – SSK	Project 677 **ST PETERSBURG** ('Lada')	1	2,700 tons	72m x 7m x 7m	Diesel-electric, 21 knots	40	2010
Submarine – SSK	Project 877/636 ('Kilo')	c. 20	3,000 tons	73m x 10m x 7m	Diesel-electric, 20 knots	55	1981

Notes:
1 Table only includes main types and focuses on operational units and/or ships under active modernisation or refit.

Construction of conventional diesel-electric submarines remains dominated by Project 636.3 'Kilo' class variants for the Pacific Fleet. *Volkhov*, the second of this batch of six boats, was commissioned in October 2020 whilst a third, *Magadan*, was launched from the Admiralty Shipyards in St. Petersburg on 26 March 2021. It appears that a third batch of the series will be built for operations in the Baltic following the announcement of an order for a new, improved boat that will be named *Petrozavodsk.* This may reflect continued concern with respect to the follow-on Project 677 'Lada' design. Although there have been reports that a sixth submarine of this type was ordered in mid-2020, the navy is still struggling to bring earlier members of the class into service.

Aircraft Carriers and Amphibious Ships: The troubled refit and modernisation of Russia's sole aircraft carrier, the Project 1143.5 *Admiral Kuznetsov*, that commenced in 2017 is still some way from completion. Current plans envisage the ship will return to the fleet in 2023 but this must be regarded as speculative given a past track record of Russian delays with similar complex projects. There have been various reports that the Russian Navy is keen to embark on construction of a replacement, nuclear-powered carrier within the lifetime of the next Russian State Armaments Programme (GPV), which is expected to cover the period 2024–33.[15]

In the interim, the start of construction on two new Project 23900 *Ivan Rogov* class amphibious assault ships – also referred to as universal landing ships – at the Zaliv shipyard in the Crimea in July 2020 will provide Russia with valuable experience in large warship construction after a gap of several

The Russian Navy's second Project 11711 *Ivan Gren* class large landing ship *Pyotr Morgunov* was delivered by the Yantar shipyard in December 2020. Two further, enlarged members of the class are currently under construction. *(Yantar)*

Russia's lead Project 22160 patrol vessel *Vasily Bykov* seen transiting the Bosphorus in December 2020. Although lightly armed in its basic configuration, the design's modular nature allows it to be equipped with containerised weapons systems in similar fashion to the Danish Stanflex concept. *(Devrim Yaylali)*

decades. The new ships are rumoured to be between 30,000 and 40,000 tons in displacement. They will be capable of deploying up to 1,000 troops and a mix of 20 helicopters and drones. Delivery is expected in the latter half of the current decade at an estimated cost of around US$1.3bn for both ships. The navy is already receiving other new amphibious ships in the form of the new Project 11711 *Ivan Gren* class large landing ships. The second of these, *Pyotr Morgunov*, was commissioned in December 2020 and has now joined her sister in Russia's Northern Fleet. Two additional, enlarged members of the class remain under construction at the Yantar shipyard in Kaliningrad that completed the first two vessels.

Surface Combatants: Russian surface combatant procurement continues to have three main strands. The largest, most complex vessels are the Project 22350 *Admiral Gorshkov* class frigates. Intended for blue water operations, these are now being built in considerable numbers after a protracted construction and trials process for the lead ship. The second class member, *Admiral Kasatonov*, was commissioned on 21 July 2020, the day after the St. Petersburg Severnaya Verf facility laid the keels of *Admiral Yumashev* and *Admiral Spiridonov*, the seventh and eighth member of the class. An additional two units were ordered in the summer of 2020, taking the class to ten ships. Russian press reports indicate that the design of a larger and improved Project 22350M frigate is close to completion, suggesting any further orders will be for the new type.

The second line of surface warship procurement relates to the Project 20380/85/86 series of *Steregushchiy* class corvettes. These are designed largely for green water operation and currently serve with the Baltic and Pacific Fleets. The last year has seen Severnaya Verf deliver *Gremyashchiy*, the first enlarged Project 20385 variant. In addition, the Amur Shipbuilding Plant in Russia's Far East has completed *Aldar Tsydenzhapov* to an improved Project 20380 design. A total of twelve Project 20380 and six Project 20385 corvettes plus the prototype Project 20386 class vessel *Merkuriy* have been ordered to date, indicating the importance of the series to the navy's future force structure.[16]

The Russian Navy is taking delivery of growing numbers of Project 12700 'Alexandrite' class MCMVs from the United Shipbuilding Corporation's Sredne-Nevsky yard. This picture shows *Pytor Ilyichev*, the sixth member of the class, during launch operations on 28 April 2021. *(United Shipbuilding Corporation)*

These larger vessels are being increasingly supplemented by a series of smaller ships largely intended for operation in coastal and confined waters but equipped with a powerful strike potential. To date, the most numerous of these are the c. 950-ton Project 21631 'Buyan M' small missile ships, which were originally designed for service in the Caspian Sea but have also been assigned to the Baltic and Black Sea Fleets. They are fitted with the UKSK S314 'Universal' VLS that can be equipped with long-range anti-shipping and land attack cruise missiles. Nine have been delivered to date by the inland Zelenodolsk yard, where three more are still to be completed. However, production is now being increasingly focused on the Project 22800 'Karakurt' design, which is broadly similar to the Project 21631 class in overall size but have a deeper draught and are more suited for open water operation. Construction of at least sixteen of the type has been assigned to five different shipyards, of which three had been delivered by the Pella shipyard at Otradnoe (downstream of St. Petersburg) as of mid-2021.

Minor Warships and Auxiliaries: Russia's surface combatant programmes are being supplemented by a wide range of other construction including patrol ships, MCMVs and logistic support vessels. Of these, the latest generation of patrol vessels such as the Project 22160 conventional and Project 23550 ice-breaking patrol ships are noteworthy in being designed to accommodate containerised strike weapons intended to further distribute the fleet's offensive capabilities. In the summer of 2020, the c. 1,800-ton *Vasily Bykov* – the lead Project 22160 patrol vessel – undertook a series of Arctic trials with the Northern Fleet that are rumoured to have included initial tests of this containerised weapons solution. *Pavel Derzhavin*, the third member of the six-strong Project 22160 series was delivered in November 2020 whilst *Ivan Papanin*, the lead member of the much larger, c. 8,500-ton Project 22550 class, is expected to commence sea trials by the end of 2022.

An interesting aspect of Russian naval procurement is the continuation of orders for traditional fibreglass-hulled MCMVs intended to operate within the minefield at a time when many navies are looking at 'mothership'/'toolbox' combinations. Four Project 12700 'Alexandrite' class MCMVs have been delivered by the St. Petersburg Sredne-Nevsky Shipyard to date, with two more – *Georgy Kurbatov* and *Pyotr Ilyichev* – launched in the last twelve months. It has been reported that up to thirty of the c. 900-ton ships might ultimately be acquired to allow operation of a modern mine countermeasures flotilla by each of the four major fleets.

Notes

1. See for example Andrew Serdy's 'What Does the Law of the Sea Say About the HMS *Defender* Incident?' posted to the Maritime Executive website – www.maritime-executive.com – on 28 June 2021. The author, professor of the Public International Law of the Sea at University of Southampton, cites Articles 17–19 of the 1982 United Nations Convention on the Law of the Sea (UNCLOS) to suggest that the British warship's transit was consistent with the provisions of international law.

2. Further information on the complex repair is contained in a Naval Group press release dated 12 April 2021, available by searching www.naval-group.com/en

3. The first four French FREMMS have two shorter Sylver A43 eight-cell modules for Aster 15 surface-to-air missiles and two longer Sylver A70 eight-cell modules that can house MdCN naval cruise missiles. It is understood the fifth and sixth FREMMS – *Bretagne* and *Normandie* – have two Sylver A50 modules and two Sylver A70 modules.

4. A detailed overview of the PO programme was posted to the subscription-based *Mer et Marine* website – www.meretmarine.com/fr – on 19 October 2020.

5. Options for three additional PPAs are believed to have lapsed. The class are being built in light, light plus and full variants, each with progressively heavier armaments.

6. A good review of the next stage of Italian naval modernisation by Luca Peruzzi, 'New Ships, Submarines And Weapon Systems For Italian Navy' was posted to the *Naval News* site – www.navalnews.com – on 23 November 2020.

7. On 12 August 2020 the Greek *Elli* class frigate *Limnos* collided with its Turkish MEKO 200TN Batch 2 counterpart *Kemalreis* whilst monitoring the activities of Turkey's research vessel *Oruç Reis* off the Dodecanese islands, highlighting the real danger for confrontation in the region.

8. For further detail, see Guy Toremans, 'The BE-NL rMCM Replacement Mine Countermeasures Capability Programme', *Maritime Security & Defence* – May 2021 (Bonn: Mittler Report, 2021), pp.38–9.

9. Jaime Karremann's marineschepen.nl website remains an excellent source of information on the Royal Netherlands Navy and its joint projects with the Belgian Naval Component.

10. For further detail see Tayfun Ozberk, 'Turkey to make its own maritime missile-launching system after sanctions interrupt Lockheed plans' posted to the *Defense News* site – defensenews.com – on 4 July 2021.

11. Devrim Yaylali's Bosphorus Naval News blog at: turkishnavy.net remains the premier English language source for Turkish Navy developments. The author thanks Mr Yaylali for his assistance in commenting on this section.

12. For an indication of the scale of the crewing crisis see Conor Gallagher, 'Defence Forces entering "irreversible decline" due to personnel crisis' posted to *The Irish Times* website – irishtimes.com – on 29 June 2021. One of the *Peacock* class – *Orla* (formerly HMS *Swift*) – is one of a number of ships reportedly out of operational service since mid-2019.

13. The Norwegian Safety Investigation Authority's report was published in May 2021 and an English-language version is available on the authority's website at: https://havarikommisjonen.no/Marine/Published-reports/2021-05-eng. The comprehensive, 213-page document – that should be mandatory reading for all warship crews – catalogues a plethora of organisational, documentation and training errors that contributed to what the Authority considered was an avoidable loss. It is interesting to note that – although extensive – most of the damage from the collision was above the waterline but the crew lacked the appropriate decision support tools to maintain effective stability.

14. The updated Swedish defence strategy has been summarised in an English language document, *Main elements of the Government bill Totalförsvaret 2021–2025*. A copy can be found at: https://www.government.se/government-policy/defence

15. A good summary of Russia's plans for its carriers and other large surface combatants was provided by Tim Ripley, 'Briefing – Russia's surface fleet ambitions: The beasts from the East', *Jane's Defence Weekly* – 5 May 2021 (Coulsdon: Jane's, 2021), pp.24–9.

16. Please see Chapter 3.2 for a detailed analysis of the *Steregushchiy* class.

2.4A FLEET REVIEW

THE SPANISH NAVY

The Armada in 2021

Author:
Alejandro Vilches

The Spanish Navy, officially called the *Armada Española*, is one of the oldest navies in the world. The *Armada* played a historic role across all the oceans and in practically all the main wars from the time of Spain's unification until the beginning of the twentieth century. However, the war with the United States in 1898 – that resulted in the loss of the last Spanish colonies in the Americas and the Pacific – and the subsequent Civil War between 1936 and 1939 saw a period of decay that lasted into the 1960s.

During the height of the Cold War, the provision of financial and materiel support from the United States first initiated by the Eisenhower Administration and efforts to build industrial ties with European countries, notably France, in the shipbuilding and naval weapons sectors marked the start of a naval revival.[1] The ultimate creation of a national naval construction industry able to build and sustain an oceanic fleet was prominent in the minds of the *Armada*'s leaders at this time. Meanwhile, the death of Spain's dictator, General Francisco Franco, in 1975 brought the opportunity for membership of NATO and the European Union. This bolstered bonds with other European countries whilst reinforcing links with the United States, which continued to maintain bases on Spanish soil.

One of the world's oldest navies, the current *Armada Española* operates some potent and successful warship designs in spite of the impact of a period of financial constraints. This 2020 image shows the F-100 type frigate *Cristobal Colon* leading the replenishment oiler *Patiño* during exercises in 2020; both designs have provided the basis for successful export programmes. *(Crown Copyright 2020)*

After some 30 years of ongoing reorganisation and periodic episodes of crisis, the creation of Navantia in 2005 achieved the consolidation of Spanish naval shipbuilding into one, focussed state-owned national champion and created the conditions for further progress.[2] Navantia and its sub-contractors have proved successful in both the domestic and international shipbuilding markets, with the overseas sale of naval vessels based on designs proven in Spanish Navy service creating an indissoluble association between the *Armada Española* and the indigenous shipbuilding sector. During the years immediately preceding and following the turn of the twenty-first century, the Spanish Government invested substantial sums in naval projects with the simultaneous aims of modernising the navy and enhancing the capabilities of the shipbuilding sector. It was during this time that the country decided to purchase the Aegis combat system, construct an advanced series of new frigates and the amphibious assault ship *Juan Carlos I*, and – perhaps most significantly – create an indigenous submarine design capability.

These ambitions of maritime advancement suffered significantly during the financial crisis that began in 2007-8 and which was to have a particu-

Spain's principal naval base at is at Rota, near Cádiz, where this photo of the now retired aircraft carrier *Dédalo* – formerly USS *Cabot* (CVL-28) – was taken in 1976. The provision of American aid from the 1950s onwards in return for basing rights at facilities such as Rota was a major catalyst for the Spanish Navy's revival after a lengthy period of decline. *(US National Archives)*

larly severe impact on the Spanish economy over the following years. The impact of this crisis is still being felt by the Spanish Navy due to delays in completing programmes that were underway at the time and the deferral of planned projects. At the same time, Spain has remained a leading global economy and an active participant in both NATO and the wider Western alliance, as evidenced for example by its participation in operations in Afghanistan and Iraq. All this suggests the financial disruption to the *Armada*'s plans is capable of being overcome in due course given the appropriate political will.

Turning to geographic considerations, one of the most significant features of Spain's location is its important strategic position astride the Strait of Gibraltar, one of the most important choke points interdicting international sea lines of communication. Moreover, the country's access to both Atlantic and Mediterranean coastlines – as well as the presence of Spanish islands in both the Atlantic and Eastern Mediterranean – supports the logistic infrastructure to project power into both seas. Accordingly, a dominant influence on naval strategy is the protection of the maritime axis that runs from the Canary Islands through the Gibraltar Strait to the Balearic Islands and has led to a need for sea-going assets that include frigates, submarines and aviation-capable ships. Another important consideration is the need to ensure the security of the African city enclaves of Ceuta and Melilla, a factor justifying the maintenance of a credible force of naval infantry (*Infanteria de Marina*) and associated amphibious shipping able to project the fleet's naval power ashore.

ORGANISATIONAL STRUCTURE AND BASES

The *Armada* is currently organised into three main departments:

- **The General Headquarters of the Navy (*Cuartel General de la Armada*):** This comprises the Admiral Chief of the General Staff of the Spanish Navy (AJEMA) and his General Staff (*Estado Mayor de la Armada*). This is located in Madrid.

- **The Fleet (*Flota*):** This encompasses the Spanish Navy's operational units, which are organised under the direction of the *Almirante de la Flota.* The *Flota* is split into a number of subordinate commands, the most important of which are:

  - The High Readiness Maritime Headquarters (*Cuartel General Marítimo de Alta Disponibilidad*). This is responsible for the major surface combatants of the Naval Action Force (*Fuerza de Acción Nava*), themselves split into two main Naval Action Groups.
  - The Maritime Action Force (*Fuerza de Acción Marítima*). This is responsible for minor naval units such as patrol and mine countermeasures vessels (MCMVs). It is largely structured on the basis of a number of regional commands centred on the main naval bases.
  - Naval Infantry Force (*Fuerza de Infantería de Marina*).
  - Airborne Forces (*Flotilla de Aeronaves*).
  - Submarine Forces (*Flotilla de Submarinos*).

- **The Force Support Organisation (*Apoyo a la Fuerza*):** This is composed of the personnel, logistical support and financial affairs bureau.

The Spanish Navy has bases distributed along both the Atlantic and Mediterranean coastlines and in the main island groupings. The principal facilities are at:

Cádiz – Rota Naval Base: Located in south-western Spain to the Atlantic side of the Strait of Gibraltar, this is the navy's main base. It is also used – and partly funded by – the US Navy, which currently has four *Arleigh Burke* (DDG-51) destroyers forward deployed at the facility. The base's facilities encompass an airfield – a major American logistical hub – that is able to handle the largest aircraft, including B-52 strategic bombers and C-5 Galaxy transporters.

Rota is the headquarters for a large number of Spanish Navy formations, including the *Flota* and several of its subordinate commands. Amongst warships based there are the fleet flagship *Juan Carlos I* and the *Galicia* class amphibious transport docks, as well as the F-80 (FFG-7 type) frigates of the 41st Escort Squadron. In addition to being headquartered at Rota, the Spanish Naval Infantry maintains a major part of its combat strength at a nearby base in San Fernando. This includes the Marine Infantry Brigade of the Tercio de Armada (TEAR) as well as the Force Protection Unit of the

The naval base and shipyard at Ferrol, one of the Spanish Navy's most important facilities and the location of the outfitting of its major warships. Amongst vessels seen in this view are four fleet replenishment vessels; the Spanish Navy's *Patiño* and *Cantabria* as well as two *Supply* class ships being completed by Navantia for the Royal Australian Navy. *(Navantia)*

The Spanish Navy's amphibious assault ship *Juan Carlos I* is home-ported at the huge Rota naval base near Cádiz, which is shared with the US Navy. This June 2020 image of the ship shows the AV-8B Plus Harriers of *9ª Escuadrilla Aeronaves* embarked. *(Crown Copyright 2020)*

Tercio del Sur. More broadly, Cádiz is the home to a conglomeration of naval shipbuilding and support facilities, including three shipbuilding and repair yards operated by Navantia.

Ferrol: Located on the Atlantic coast of north-west Spain, Ferrol is another principal Spanish naval base. Located in proximity to Navantia's Ferrol/Fene shipyards, it is home to the 31st Escort Squadron of F-100 type Aegis-equipped frigates and a number of other vessels. The Spanish Naval Infantry's Tercio del Norte unit is based here.

Cartagena: Cartagena is the main Spanish naval facility on the Mediterranean coast. It is the headquarters of the *Flota*'s Submarine and Maritime Action Force commands. The Navantia yard responsible for submarine construction is located here. Units based at Cartagena include the current two operational submarines, the *Segura* class MCMVs and many patrol vessels and auxiliaries. The Tercio de Levante of the Naval Infantry is also based at Cartagena.

Canary Islands – Las Palmas de Gran Canaria: This is the base for a number of offshore patrol vessels of the *buque de acción marítima* (BAM) type and a small Naval Infantry support unit. It is also very common for other units of the navy to be deployed to the archipelago as forward units.

Other: In addition to these major facilities, the *Armada* also maintains a network of minor bases around the Iberian Peninsula, as well as in the Balearics and North African enclaves. These can be used to support the deployment of appropriate units as the situation warrants.

NAVAL UNITS

The Spanish Navy is currently comprised of a mixture of units dating from the Cold War era supplemented by ships of more recent construction. A list of major units is contained in Table 2.4A.1, with further detail provided below.

Aircraft Carrier: The navy's most significant unit is the large-deck amphibious assault ship *Juan Carlos I*, which currently performs both the amphibious and aircraft carrier roles. Originally referred to as a *buque de proyección estratégica* (BPE) or strategic projection ship, she is now commonly referred to as a *buque* *anfibio portaeronaves* or amphibious aircraft carrier. She effectively replaces the previous aircraft carrier *Príncipe de Asturias*, which was based on the US Navy's sea control ship concept designed by Gibbs & Cox but decommissioned in 2013.

With a full load displacement of 27,000 tons, *Juan Carlos I* is able to deploy Harrier jump jets and all the helicopters currently held in the Spanish military's inventory and has also proved capable of operating the US Marine Corps' MV-22B Osprey in shipboard trials. Up to thirty aircraft of various types can be embarked. Her ski-jump equipped flight deck is laid out with six refuelling and rearming points and served by two aircraft lifts. Her core crew only amounts to c. 250 but she can embark a significant complement of aviation personnel and a substantial force of naval infantry, as well as supporting equipment stowed in two flexible hangar/vehicle decks. Her aft well deck can carry four LCM mechanised landing craft or two Landing Craft Air Cushion (LCAC).[3]

The *Juan Carlos I* design – combining aviation, amphibious and command functions – has been a major success for Spanish shipbuilding. Australia acquired two of the type as its *Canberra* class whilst Turkey has purchased the design and associated technical assistance to construct *Anadolu* at the Sedef yard in Istanbul. A second Turkish vessel may be built in due course, whilst the design has also been proposed for India's amphibious assault ship programme.[4]

Juan Carlos I is sized to operate the F-35B Lightning II strike fighter but some modifications would be needed before any shipboard trials could be performed. To date, no work has been carried out in this regard.

Surface Combatants: The surface component of the Spanish Navy is based on its squadrons of frigates. During most of the Cold War, Spain's force of surface combatants was heavily focused on units transferred from the United States but this subsequently progressed to the construction of American designs in Spanish shipyards. The most recent of these are the *Santa María* class, which were based on the famous *Oliver Hazard Perry* (FFG-7) design. Six of these vessels were completed between 1986 and 1994 and remain in service pending replacement by the new F-110 design. All are based at Rota. Their principal armament is a single Mk 13 launcher for Standard SM-1 surface-to-air and Harpoon surface-

Table 2.4A.1 SPANISH NAVY FLEET COMPOSITION – MID 2021

TYPE	CLASS	NUMBER	DATE[1]	SHIPS
Principal Surface Escorts (11)				
Frigate (FFG)	**BAZÁN** (F-100)	5	2002–2012	*Álvaro de Bazán* (F101); *Almirante Juan de Borbón* (F102); *Blas de Lezo* (F103); *Méndez Núñez* (F104); *Cristóbal Colón* (F105)
Frigate (FFG)	**SANTA MARÍA** (F-80)	6	1986–1994	*Santa María* (F81); *Victoria* (F82); *Numancia* (F83); *Reina Sofía* (F84); *Navarra* (F85); *Canarias* (F86)
Submarines (2)				
Patrol Submarine (SSK)	**GALERNA** (S-70)	2	1983–1986	*Galerna* (S71); *Tramontana* (S74)
Amphibious Ships (4)				
Amphibious Assault Ship (LHD)	**JUAN CARLOS I**	1	2010	*Juan Carlos I* (L61)
Amphibious Transport Dock (LSD)	**GALICIA**	2	1998–2000	*Galicia* (L51); *Castilla* (L52)
Logistical Transport	**YSABEL**	1	2021	*Ysabel* (A06)
Replenishment Vessels (2)				
Replenishment Oiler (AOR)	**PATIÑO**	1	1995	*Patiño* (A14)
Replenishment Oiler (AOR)	**CANTABRIA**	1	2010	*Cantabria* (A16)
Mine Countermeasures Vessels (6)				
Minehunter – MCMV	**SEGURA**	6	1999–2005	*Segura* (M31); *Sella* (M32); *Tambre* (M33); *Segura* (M34); *Duero* (M35); *Tajo* (M36)
Offshore Patrol Vessels (15)				
Offshore Patrol Vessel (OPV)	**METEORO** (BAM)	6	2011–2019	*Meteoro* (P41); *Rayo* (P42); *Relámpago* (P43); *Tornado* (P44); *Audaz* (P45); *Furor* (P46)
Offshore Patrol Vessel (OPV)	**DESCUBIERTA**	2	1980	*Infanta Elena* (P76); *Infanta Cristina* (P77)
Offshore Patrol Vessel (OPV)	**SERVIOLA**	4	1991–1992	*Serviola* (P71); *Centinela* (P72); *Vigia* (P73); *Atalaya* (P74)
Offshore Patrol Vessel (OPV)	**CHILREU**	3	1997–2004	*Alborán* (P62); *Arnomendi* (P63); *Tarifa* (P64)

Notes:

1. Date refers to the delivery date(s) of ships in the class remaining in service.
2. Other ships in service include eight smaller patrol boats; eight hydrographic/research vessels and launches; three salvage tugs; a flotilla of training vessels (including the sail training vessel *Juan Sebastián de Elcano*); the intelligence-gathering vessel *Alerta;* the submarine rescue ship *Neptuno*; and twelve LCMs.

The lead F-100 class frigate, *Álvaro de Bazán*, pictured operating in the Mediterranean in 2016. Consolidating many destroyer-like capabilities in a frigate-sized hull, the F-100 class design is focused on the air-defence role and was the first to incorporate the Aegis combat system in such a small ship. *(US Navy)*

The *Galerna* class submarines have remained in service longer than anticipated due to delays in the replacement S-80 Plus programme. Seen here on completion of a refit in 2013, *Mistral* was finally decommissioned in February 2021, leaving just two submarines in Spanish Navy service. *(Navantia)*

to-surface missiles, supplemented by Mk 32 anti-submarine torpedo launchers, a 76mm Oto Melara (Leonardo) gun and the Meroka close-in weapons system. The class can also operate up to two SH-60B Seahawk helicopters.

When the Spanish Navy commenced the programme for its following class of frigates, they decided to install the Aegis combat system and vertical launch systems (VLS) in the new vessels. The resulting F-100 type frigates of the *Álvaro de Bazán* class were built by Navantia with considerable support from US industry. Consolidating many destroyer-type capabilities in a frigate-sized hull, the resulting ships placed the Spanish shipbuilding sector at the forward edge of the naval construction sector's technological capabilities. Notably, it was the first time that the Aegis system had been installed in such a small vessel.

Displacing around 6,250 tons, the F-100 class frigates are armed with a 48-cell Mk 41 VLS for Standard SM-2 MR and Evolved Sea Sparrow missiles, two quad Harpoon surface-to-surface missile launchers, a single Mk 45 127mm gun and Mk 32 torpedo launchers. They can also operate a single SH-60B Seahawk. Four members of the class were completed between 2002 and 2006. An additional, fifth unit was delivered to the *Armada* in 2012 but a planned sixth unit was never ordered. The total cost of the original quartet was less than €2bn (US$2.5bn) but the unit cost increased to more than €800m (US$1bn) for the fifth ship. Nevertheless, the class represented another major success for Spanish shipbuilding. Australia's three *Hobart* class destroyers were essentially built to the same design whilst Norway's *Fridtjof Nansen* class frigates are a derivative.

Submarines: The submarine flotilla is currently the most problematic issue facing the fleet. Traditionally, a positive industrial relationship with France's shipbuilding sector had resulted in Paris providing the design and technological expertise required to support the construction of submarines at Cartagena. This partnership had extended beyond the Spanish Navy's needs to encompass joint construction of submarines for export, notably the 'Scorpène' type boats sold to Chile and Malaysia. However, during the good years prior to the economic crash, the Spanish Government decided to invest in a very ambitious programme to be able to design and build its own submarines

Amphibious shipping is an important component of the modern *Armada Española*. Designed in collaboration with the Royal Netherlands Navy and similar to HNLMS *Rotterdam*, the two amphibious transport docks of the *Galicia* class were commissioned in 1998 and 2000. This is *Galicia* in October 2019. *(US Navy)*

without French assistance. In one sense, this decision was well-conceived given the success of Spanish shipbuilding in exporting other warship types. However, implementation of the programme – that was envisaged as ultimately providing eight units – was to prove problematic. This has resulted in a large decline in submarine numbers that will only be partly reversed when the new programme is realised.

Currently the Spanish Navy only has two submarines in commission, both French-designed but Spanish-built boats of the *Galerna* – French *Agosta* – class. Displacing c. 1,800 tons submerged, they are armed with four 533mm torpedo tubes. They are equipped with French-made torpedoes – the F-17 and L-5. The survivors of a class of four, the pair were commissioned in 1983 and 1985 and have been modernised to extend their service lives to the maximum extent possible until their replacements enter service. At the moment *Galerna* herself is undergoing heavy overhaul, meaning that just one boat is operational.

Amphibious Ships: In addition to the previously mentioned *Juan Carlos I*, the fleet operates two additional, versatile amphibious units of the *Galicia* class amphibious transport dock (LPD) type. Designed in collaboration with the Royal Netherlands Navy, these ships were commissioned in 1998 and 2000 and are based at Rota. They displace around 13,000 tons and have an overall length of 160m. They are able to provide accommodation for up to c. 600 naval infantry and their equipment, as well as operate helicopters and LCMs in support of both

The replenishment oiler *Cantabria* is an improved, double-hulled development of the older *Patiño*. This photograph shows her in a three-ship replenishment with (left to right) the Canadian *Halifax* class frigate *Winnipeg*, the German F124 class *Hamburg* and the lead member of the Spanish *Santa María* class. *(Navantia)*

The modular *buque de acción marítima* (BAM) offshore patrol vessel *Meteoro* seen during her final trials in July 2011. Six of the class have been delivered to date. *(Navantia)*

amphibious and non-combat operations. A typical load-out would encompass four medium helicopters and four LCMs in the well deck.

Amphibious assaults would be supported by the fleet's total of twelve LCM-1E landing craft, which were both designed and constructed in Spain. The last of these was delivered in 2008. The Royal Australian Navy also operates the same type from its own amphibious ships.

In June 2021, the *Armada* updated its logistical transport capabilities with the delivery of the second-hand Ro-Ro vessel *Galicia*, which has now taken the name *Ysabel.*

Support Ships: As is the case for any other navy maintaining naval projection capacities, the Spanish Navy operates support and logistics vessels to allow its fleet to be deployed to areas at distance from its home ports. The *Armada* currently has two vessels capable of conducting replenishment of diesel, aviation fuel, fresh water, ammunition and other stores at sea. The replenishment oiler (AOR) *Patiño* – delivered in 1995 – was subsequently joined by the slightly larger and more modern *Cantabria*, which was commissioned in 2010.[5] The latter design is another that has been sold to Australia, which received two *Supply* class AORs built by Navantia at the Ferrol/Fene yards, in 2020 and 2021.

Mine Countermeasures Vessels: Six modern units of the *Segura* class are currently in service. Based on the British Royal Navy's *Sandown* class design, they have a distinct Spanish equipment fit that includes the Raytheon SQQ-32 variable-depth sonar, Pluto Plus remotely-operated vehicles and Kongsberg's Minesniper mine-disposal system. All were delivered between 1999 and 2005. Navantia undertook substantial investments in infrastructure, tools and training to produce the class but programme numbers were curtailed and no exports were achieved.

In August 2019, the fourth unit of the class, *Turia* (M34), grounded in bad weather off Spain's Mediterranean coast close to Murcia whilst undertaking salvage operations following the crash of a Spanish Air Force jet trainer with the loss of its pilot. There was heavy damage to the ship's fibreglass hull and there were rumours that the Spanish Ministry of Defence might dispose of the vessel. However, after detailed studies, it was ultimately determined to repair the vessel at a cost of some €25m

The Spanish Navy SH-60B Seahawk helicopters of *10ª Escuadrilla Aeronaves* provide the core of its airborne sea control capabilities but are now in need of urgent replacement. *(Crown Copyright 2020)*

(c. US$31m). The work will be carried out by Navantia and the ship should return to service by the end of 2022.

Minor Units: The Spanish Navy currently operates a varied collection of offshore patrol vessels (OPVs) of various ages and sizes. The *Armada*'s long term goal has been to replace many of these ships with the modular *buque de acción marítima* (BAM) design. The intention was to construct a series of largely identical vessels that could be easily built to perform a particular mission by incorporating the relevant equipment and systems required for the role. As well as constabulary duties, it was envisaged that BAMs could be constructed to meet other fleet requirements, such as hydrography or salvage and rescue.

An initial order for four BAM-type OPVs – the *Meteoro* class – was placed in 2006, with delivery taking place during 2011–12. The financial crisis meant that it was not until 2014 that a further batch of just two units was authorised, with the two resultant OPVs delivered during 2018–19. Although relatively lightly armed with a 76mm main gun, they are large, flexible ships with good command and communications facilities equally capable of extended EEZ patrols or overseas stabilisation missions. However, the slow progress achieved with the programme means that many old patrol vessels remain in service and their replacement is becoming increasingly pressing.

To date, no firm orders for other BAM variants have been placed. However, in late 2020, the Spanish Government authorised a seventh unit at a reported cost of some €183m (c. US$229m) that will be configured as a submarine search and rescue vessel to replace the current elderly *Neptuno*. The so-called BAM-IS is seemingly based on the original BAM hull form but has a radically different appearance from the previous OPVs. There are also near-term plans for a further vessel that will be focused on electronic intelligence (ELINT). Interestingly, the Spanish Navy currently operates the former East German intelligence ship *Alerta*, which was purchased from Germany after reunification and supposedly remains equipped with intelligence gathering equipment of Soviet origin. This is useful for operations against Mediterranean countries along the North African coast that use similar communications equipment.

Amongst other navy vessels are various research and hydrographic ships, as well as tugs, training ships and launches.

NAVAL AIR ARM

Headquartered at Rota, the *Armada*'s small Naval Air Flotilla comprises a total of seven squadrons of fixed-wing aircraft, helicopters and unmanned aerial vehicles (UAVs).

The most important fixed wing component is arguably *9ª Escuadrilla Aeronaves*. This comprises twelve AV-8B Plus Harrier aircraft and a single TAV-8B trainer. These aircraft currently provide the embarked fixed-wing component of the amphibious assault carrier *Juan Carlos I*. They will reach the end of their operational lives before the end of the decade and a decision with respect to their replacement needs to be taken urgently. The only feasible option is the F-35B variant of the Lightning II but this is a major political decision involving considerable expense that will not be taken easily. The navy will lose much of its capacity to mount expeditionary operations, particularly across the North African region, if its embarked fixed-wing capability is lost.

Many of the fleet's helicopter squadrons are also in need of modernisation. It was originally intended to replace the twelve SH-60B Seahawk sea control helicopters of *10ª Escuadrilla Aeronaves* with the NHF variant of the European NH90 helicopter. However, these plans were postponed due to the financial crisis and the navy is reported to be seeking the acquisition of the MH-60R Seahawk variant as an interim replacement. Meanwhile the remaining veteran Sikorsky SH-3D Sea Kings of *5ª Escuadrilla Aeronaves*, which are used for troop transport, are being steadily replaced by second-hand SH-60F Seahawks.[6] *3ª Escuadrilla Aeronaves* – the other naval transport squadron – operates seven AB-212 light helicopters in the utility role. *6ª Escuadrilla Aeronaves* is largely responsible for training with its Hughes H-500 helicopters.

The navy maintains some surveillance capacity through the land-based Cessna Citation jets of *4ª Escuadrilla Aeronaves*. These are supplemented by the handful of operational P-3B Orion maritime patrol aircraft maintained by the Spanish Air Force. The P-8 Poseidon or modernised Atlantique 2 are the only currently available aviation platforms that could replace this capability but it seems that Spain will need to make do with the less well-suited Airbus CN-235, which is manufactured indigenously. Meanwhile, the Scan Eagle UAVs of *11ª Escuadrilla Aérea* provide the navy with a useful embarked surveillance system.

NAVAL INFANTRY

The Spanish Navy's Naval Infantry, established in 1537, is considered to be the oldest marine corps in the world. Comprising between 5,000 and 6,000 personnel, it is essentially split between the Marine Infantry Brigade and supporting elements of the TEAR, the regionally-based Tercios of the Force Protection Unit and the Cartagena-based Naval Special Warfare Force. The all-arms Marine Infantry Brigade gives the navy a significant 'in house' expeditionary capability, providing the means of protecting Spain's territories along the North African coastline should the need arise.

A graphic of the new F110 *Boniface* class frigate. Navantia will complete five of these ships over the next decade. *(Navantia)*

NEW CONSTRUCTION PROGRAMMES

The financial crisis of the last decade has severely impacted nearly all the navy's programmes, some of which have been delayed beyond reasonable limits. A number of new projects now have more momentum, the most significant being listed below.

F-110 Frigates: The replacement for the *Santa María* class, these are derived from the previous F-100 design and will also use Aegis combat system technology. In contrast to the F-100, however, they will be a general-purpose design with an emphasis towards anti-submarine warfare operations. Published data suggests they will displace 6,100 tons, have a length of 145m and a beam of 18m.

Construction of the first of class, *Boniface*, is expected to commence at Navantia Ferrol in 2022, with the preliminary design review completed in May 2021. Five ships will be delivered at annual intervals through to the early 2030s. Total programme expense is estimated to be c. €4.3bn (c. US$5.2bn), with production cost excluding research and development in the region of €740m (c. US$900m) per unit. The reason for their reduced number compared with the ships they will replace is based on their enhanced capabilities but funding – or lack thereof – is probably the real reason. Navantia hope that the programme will emulate the success of the previous F-100 design and a variant has already been unsuccessfully offered to meet the Hellenic Navy's new frigate requirement. Their impact on the export markets may, however, be influenced by US control over the American-supplied technology they incorporate.

S-80 Plus Submarines: Originally approved nearly two decades ago, what was originally the S-80 programme is probably Spain's best-known naval construction project due to the numerous problems that it has experienced. Of these, the most notable was an error in weight calculations that could have caused the boats problems with surfacing. General Dynamics' Electric Boat division was ultimately brought into help Navantia redesign the submarines, resulting in an increase in length that has been reflected in the new 'Plus' designation.

The programme now appears to be largely back on track, with the lead submarine – *Isaac Peral* – launched in May 2021 prior to a lengthy period of pre-delivery tests and trials. Three further boats are at various stages of construction. The new class will be amongst the largest and – in theory – the most advanced conventional submarines in the world, the latter by virtue of their ethanol-based air independent propulsion (AIP) plants. However, this technology has suffered even greater delays than the submarines themselves and is not expected to be

The lead S-80 Plus submarine *Isaac Peral* pictured shortly after being rolled out of her building hall and subsequent launch in May 2021. The S-80 submarine programme has been beset by problems but is now seemingly back on track. *(Navantia)*

installed 'on build' in the first two units. As previously mentioned the considerable delays experienced by the project have also had an impact on existing operational submarine strength, even though members of the legacy *Galerna* class have been subject to additional dockings and refits to keep at least some boats operational.

The programme's growing cost has also had an impact on the Spanish Navy's overall funding over the last decade, impacting the money available for other platforms. The original budget allocated to the project was c. €1.8bn (c. US$2.25bn) but this has progressively increased to around €3.9bn (c. US$4.8bn) or approximately US$1.2bn per unit. This is a substantial amount compared with rival designs from, say, France or Germany, but would probably not be reflected in any subsequent boats. As such, the initial cost might be justifiable from an industrial perspective if exports can be achieved. Navantia was unsuccessful promoting the design for Australia's new submarine programme despite its track record with the Royal Australian Navy due to the design's problems but – now progress is being achieved – may be better placed in its bid for India's new P75I project.

With a submerged displacement of nearly 3,000 tons and an overall length of 81m, the S-80 Plus has six torpedo tubes and can stow a total of eighten weapons. In contrast to the French outfit of weaponry on the existing Spanish submarines, it will be equipped with German DM2A4 SeaHake heavyweight torpedoes and US Sub-Harpoon surface-to-surface missiles and has the capacity to fire the

Tomahawk cruise missile.[7] The submarine's fire control system is also of US origin.

BAM OPVs: The BAM design has proved to be a versatile and comparatively affordable platform that, as indicated earlier, can be re-configured to perform a variety of roles. The imminent start of construction of the seventh BAM-IS variant suggests that the design has continued relevance for the *Armada* in a number of specialised missions. However, Navantia's decision to join Fincantieri and Naval Group in the European Union-supported European Patrol Corvette project may well direct any future orders for constabulary vessels in that direction. More broadly, whilst Navantia's achievements in exporting OPVs have, perhaps, not quite met the success it has gained with higher-end warships, sales of patrol vessels to Venezuela and Morocco, as well as of corvettes to Saudi Arabia, have all been valuable in maintaining shipbuilding activity.

CONCLUSION

As has been the case for many other Western navies, the *Armada* has been shrinking in overall size on the basis that the introduction of more capable vessels incorporating new technological advances is adequate compensation for a decline in numbers. The impact of the significant financial problems of the recent past has also been particularly detrimental to the navy's capacity. Whilst government budgetary restrictions in a time of financial crisis have not been unreasonable, there has inevitably been an effect on the navy's ability to meet the mission requirements it is expected to perform.

The importance attached to the Spanish naval construction sector has seemingly partly mitigated these problems, as evidenced by the support provided to completing the S-80 Plus programme. The new frigate programme is also now underway. This possibly leaves naval aviation as the most substantial and pressing problem, with new fixed wing aircraft and helicopters required as existing assets wear out. A failure to replace these aircraft would severely impact the navy's ability to respond to a future crisis, particularly in the amphibious sphere.

As governments prepare for the financial uncertainties of the post-COVID period, the Spanish Navy's future direction remains in question.

The F-100 *Álvaro de Bazán* class frigate *Cristobal Colon* makes an impressive sight in this late 2020 photograph. The Spanish Navy has made significant strides in operating modern, technologically-advanced vessels with the support of domestic naval shipbuilder Navantia but further investment is needed after a period of financial austerity. *(Crown Copyright 2020)*

Notes

1. American support for the Franco regime resulted from the Madrid Pacts of 1953, which provided for the provision of financial and military assistance to Spain in return for the location of American bases on Spanish soil.

2. Navantia traces its origins to the incorporation of Empresa Nacional Bazan de Construcciones Navales Military SA in 1947 as the Spanish state-owned vehicle for naval shipbuilding. This subsequently merged with the Astilleros Españoles SA merchant shipbuilder to become IZAR Construcciones Navales in 2000. A further reorganisation saw the largely naval business of the group separated out into Navantia in 2005.

3. For further details of the *Juan Carlos I* design, see the editor's 'Modern European Amphibious Assault Ships' in John Jordan (editor), *Warship 2011* (London: Conway, 2011), pp.80–93.

4. The *Juan Carlos I* design was also under serious consideration when Russia ultimately decided to acquire two *Mistral* class amphibious assault ships from France.

5. *Cantabria* differs from *Patiño* in being an up-to-date, double-hulled design complying with the latest environmental standards. Although somewhat larger than her predecessor, the space trade-offs associated with her double hull mean that her liquid stores capacity is not significantly different from her predecessor's. She is, however, specially equipped to handle containerised stores and benefits from being equipped with a modern combat management system.

6. The *Armada* is also due to receive 'marinised' NH90 TTH variant transport helicopters under a contract agreed in 2018 but it is not clear to what extent this has been superseded by the SH-60F purchases.

7. Plans to acquire submarine-launched Tomahawk cruise missiles have been cancelled on financial grounds.

2.4B FLEET REVIEW

Author:
Richard Beedall

THE ROYAL NAVY

Rebuilding Britannia

Since the end of the Second World War in 1945, the story of the United Kingdom's Royal Navy (RN) has been one of gradual decline interrupted by sudden deep cuts; the last such instance being announced in 2010.[1] The service has since suffered from a decade of negative commentary, often being unfavourably compared against the capabilities of the French *Marine Nationale.*

There was thus scepticism in the 'Senior Service' when, on 19 November 2020, the United Kingdom's Prime Minister, Boris Johnson, made a speech pledging to 'restore Britain's position as the foremost naval power in Europe … If there was one policy which strengthens the UK in every possible sense, it is building more ships for the Royal Navy.'[2] However, it was simultaneously announced that additional funding would be added to the Defence Budget; when combined with previous commitments this amounted to an extra of just over £24bn (c. US$34bn) over four years. UK defence spending will thus increase from c. £42.5bn (c. US$58bn) in 2020/21 to c. £48bn (c. US$66bn) in real terms in 2024/25 (c. 2.2 percent of GDP). Most of the additional expenditure is allocated to capital spending (i.e. new equipment), which will rise substantially in real terms.

The Royal Navy is given a prominent role in the British Government's Integrated Review of Security, Defence, Development and Foreign Policy. This image shows the new aircraft carrier *Queen Elizabeth* and the Type 23 frigate *Northumberland* being refuelled by the fleet tanker *Tideforce* in the course of the 'Westlant 19' deployment. *(Crown Copyright 2019)*

THE INTEGRATED REVIEW 2021

On 16 March 2021, the UK government published '*Global Britain in a competitive age: The Integrated Review of Security, Defence, Development and Foreign Policy*'.[3] This states that Russia poses the most acute threat to the UK's security, but also expresses concerns about an increasingly assertive China. As such it proposes deeper UK engagement in the Indo-Pacific region – which has had a minimal UK military presence for over 50 years. A week later the Ministry of Defence (MoD) issued a Command Paper '*Defence in a Competitive Age*' which described in more detail (although with important gaps) how the UK's armed forces will change as a result.[4] The two documents are hereafter referred to as 'IR2021'.[5]

IR2021 more than spends the additional funding announced in 2020. It is soaked up by the creation of new capabilities such as a National Cyber Force and a Space Command; increased Research and Development; and filling a pre-existing 'black hole' of unfunded projects in the defence equipment budget. Cuts to the three traditional military arms are thus required. However, unlike the British Army and the Royal Air Force (RAF), the Royal Navy escapes quite lightly.

The RN will decommission two of its thirteen remaining Type 23 frigates – *Monmouth* and *Montrose* – in 2021–2, some four to five years earlier than planned. This will leave just seventeen escorts in service. IR2021 also states that all thirteen remaining 'Hunt' and *Sandown* class mine countermeasure vessels will decommission over the next decade but provides few details as to how they will be replaced beyond confirming that '…this will include delivering a safer, faster and automated Mine Hunting Capability (MHC) in partnership with France'.

Looking forward, the Command Paper refers to 'Growing maritime capabilities [which] will enable the projection of power further afield and to conduct operations from increased range.' A significant naval construction programme is described, with the First Sea Lord (1SL), Admiral Radakin, claiming that the Royal Navy will have 50 percent more tonnage of warships in 2030 than it had in 2015.[6] However, previous defence reviews have made immediate cuts to the RN whilst never delivering many of the promised new ships, leaving some grounds for continued scepticism. It takes so long and costs so much to regain capabilities that the RN is still suffering from the legacy of the 2010 cuts.

ROLE OF THE ROYAL NAVY

Official thinking about the role of the Royal Navy has evolved and expanded in ambition since the last defence review in 2015, when the UK was still part of the European Union.[7] The threats were then less obvious and any increase in defence spending difficult to justify.

IR2021 essentially implies five core capabilities for the RN:

- **Maintaining the UK's Continuous at Sea Deterrent (CASD):** There is always a Royal Navy ballistic missile submarine (SSBN) on patrol, armed with strategic nuclear missiles.
- **North Atlantic:** Protecting the security of the British Isles and surrounding waters.
- **UK Carrier Strike Group:** A powerful sovereign military capability available for worldwide deployment.
- **Littoral Strike:** High readiness amphibious capabilities based on specialist ships and the Royal Marines Commando Force.
- **Forward deployed naval assets:** A small number of warships strategically based around the globe.

The next sections consider each of these in turn.

Table 2.4B.1: ROYAL NAVY FORCE DEVELOPMENT

PROGRESSIVE REVISION OF FUTURE FORCE PLANS IN DEFENCE REVIEWS 2010-2021: SELECTED SHIP TYPES

Ship Type Date of Review[1]	Future Force 2020 2010 (SDSR)	Joint Force 2025 2015 (SDSR)	Integrated Force 2030 2021 (IR)	Notes as of 2030
Aircraft Carrier – QEC	1+1[2]	2	2	
Destroyer – Type 45	6	6	6	Replacement Type 83 announced
Frigate – Type 23	13	8	5	Being replaced by Type 26
Frigate – Type 26	–	–	3	5 more ships under construction
Frigate – Type 31	–	5	5	Up to 5 Type 32 frigates planned
SSBN – VANGUARD	4	4	3	Being replaced by *Dreadnought* class
SSBN – DREADNOUGHT	–	–	1[3]	3 more boats under construction
SSN – TRAFALGAR	2	–	–	
SSN – ASTUTE	5	7	7	
LPD – ALBION	1+1[2]	1+1[2]	1+1[2]	Being replaced by up to 6 MRSS
LSD – 'BAY'	2+1[2]	3	3	As above
MCMVs	14	12	–	Being replaced by the autonomous MHC capability
OPVs	4	Up to 6	5	River Class Batch 2s
Fleet Tankers	Not stated[4]	6	6	
Fleet Solid Support Ships	Not stated[4]	3	3	

Notes:

1. In broad terms, each recent defence review has set out the navy force structure planned 10 years in the future. The abbreviation in brackets refers to the defence review setting out the future plan referenced; in some cases numbers have been inferred rather than specifically stated. SDSR is Strategic Defence & Security Review; IR is Integrated Review. The last-mentioned is particularly vague in setting out precise future force numbers.
2. The 'plus' figure refers to ships held in extended reserve.
3. The *Dreadnought* class submarines will enter service from 'the early 2030s'.
4. SDSR 2010 stated that there would be 'A fleet of resupply and refuelling vessels scaled to meet the Royal Navy's requirements'.

NUCLEAR DETERRENT

The Royal Navy has provided a CASD since May 1969 and will continue to do so until at least the 2060s. The current *Vanguard* class SSBNs entered service in the 1990s and their service life is being stretched as the replacement *Dreadnought* class will not be available to take over the task until the early 2030s.

Construction of the first of class, *Dreadnought*, began in 2016 and hull segments are now being joined together in the Devonshire construction shed at BAE Systems Submarines, Barrow-in-Furness. Despite reports of of potential risks relating to both the PWR3 nuclear reactor (provided by Rolls-Royce) and the Common Missile Compartment (provided by General Dynamics Electric Boat in the USA), the MoD is insistent that work on *Dreadnought* is progressing on schedule and within budget; as such her sea trials should commence in 2028.[8] Construction of the second boat, *Valiant*, began in 2019. *Warspite* should follow in 2022, and finally *King George VI* in 2024–5.

A surprise announcement in IR2021 was the reversal of a gradual reduction in the number of nuclear warheads held by the UK from 225 to 180 by the mid-2020s, of which only 120 were to be

An unusual view of the *Sandown* class MCMV *Grimsby* taken as she was being transferred back to the water from Babcock's maintenance hall at Rosyth in 2012. The Royal Navy's manned MCMVs are to be replaced by unmanned and autonomous systems over the next decade. *(Crown Copyright 2012)*

operational. Instead the total stockpile will increase to 260. Whilst the revised number is still lower than any of the other four recognised nuclear powers (the next lowest is France with 290), the decision is controversial. The increase is possibly because a warhead replacement programme might temporarily reduce the percentage of the total that is operational. Additionally, the UK may be seeking the capability to target and credibly deter more than one potential enemy.

The UK will also no longer declare how many operational Trident II missiles its SSBNs carry – the *Vanguard* class currently patrol with a load of only eight missiles in their sixteen tubes. Previous government announcements said that the *Dreadnought* class would also carry eight missiles, although the submarines will be fitted with twelve tubes.

Argyll, the oldest Type 23 frigate remaining in Royal Navy service, pictured entering Portsmouth Harbour in May 2021. Two of her sisters will be retired early as a result of the Integrated Review but an expansion of maritime capabilities – supported by ambition construction plans – is anticipated in the longer term. *(Derek Fox)*

The strategic submarine *Vigilan*t pictured transiting the Clyde under the watchful gaze of the patrol boat *Tracker* in March 2019. Along with her sisters in the *Vanguard* class, she will be replaced by the new *Dreadnought* class boats from the early 2030s. *(Crown Copyright 2019)*

NORTH ATLANTIC AND THE BRITISH ISLES

From the end of the Cold War in 1990 until as recently as 2014 there seemed to be no significant military threat to the British Isles, resulting in many UK defence capabilities being reduced or 'gapped'. The resurgence in Russian military activity in UK waters thereafter caught out both the Royal Navy and the UK government. There has since been some frantic recovering of capabilities; most prominently an urgent order in 2016 for nine Boeing MRA1 (P-8A) Poseidon maritime patrol aircraft. Five were in RAF service as of mid-2021 and the remainder should follow by January 2022.

IR2021 acknowledges that the seas around the UK are populated by critical infrastructure such as underwater pipelines and communication cables. It is believed that the Russians are actively trying to tap these or preparing measures to cut these links in the event of a crisis, but the RN has had only limited countermeasure capabilities since the decommissioning of the diving support ship *Challenger* in 1990. Despite the vague name, the purpose of the new Multi Role Ocean Surveillance Ship announced by IR2021 is undoubtedly to monitor and defend underwater infrastructure.

The UK's departure from the European Union on 31 January 2020 places additional pressures on the challenge of policing an exclusive economic zone (EEZ) covering 298,718 square miles. The RN has just two 'River' class Batch 1 patrol vessels – *Mersey* and *Tyne* – assigned to this task, supplemented by a third vessel – *Severn* – which also has training duties. Whilst there are also several smaller ships from the Border Force and Marine Scotland that also perform surveillance operations, the density of patrols is very low.

The Royal Navy is required to provide an on-call frigate/destroyer for urgent operations around the British Isles, designated the Fleet Ready Escort (FRE). However, because of the small number of frigates and destroyers available, the warship designated as FRE is often simultaneously assigned other tasks such as Towed Array Patrol Ship (escorting SSBNs entering and leaving Faslane), training duties and even North Atlantic/Northern Europe deployments. As a result, patrol vessels are frequently utilised for FRE-type duties such as accompanying Russian warships through UK waters.

Queen Elizabeth pictured in the course of training operations in October 2020 in advance of the CSG21 global deployment. *(Crown Copyright 2020)*

CARRIER STRIKE GROUP

The first operational deployment (CSG21) of the UK Carrier Strike Group began on 22 May 2021, marking a major milestone for a programme that dates back to the 1990s and which was often in danger of cancellation due to cost.[9] CSG21 will spend most of the 28-week deployment in the Mediterranean, Indian Ocean and Pacific, including visits to Oman, India, Singapore, Korea and Japan. CSG21 consists of:

- The aircraft carrier HMS *Queen Elizabeth*, the RN's fleet flagship.
- The Type 45 destroyers *Diamond* and *Defender*.
- The Type 23 frigates *Kent* and *Richmond*.
- An unnamed *Astute*-class nuclear submarine.
- The stores ship RFA *Fort Victoria*.
- The fleet tanker RFA *Tidespring*.

The US Navy destroyer *The Sullivans* (DDG-68) and the Dutch frigate *Evertsen* are also integrated in to the CSG. Other Royal Navy ships will join for parts of the deployment.

Although smaller than hoped when planning for CSG21 began, *Queen Elizabeth*'s air group is still the largest embarked by a RN carrier for 39 years, being composed of:

- Eighteen F-35B Lightning strike fighters from the RAF's 617 Squadron and US Marine Corps VMFA-211.
- Seven Merlin HM 2 helicopters of 820 Naval Air Squadron. Whilst primarily intended for anti-submarine warfare, three helicopters are fitted with the new Crowsnest radar system for the Airborne Surveillance and Control (ASaC) role.

In addition:

- Three Merlin MC 4 helicopters of 845 NAS are based on *Fort Victoria*. These are to be used for vertical replenishment, maritime intra-theatre lift and littoral strike (landing Royal Marines).
- Four Wildcat HMA 2 helicopters from 815 NAS form flights for the Type 45 and 23 escorts.

The authorised ship's company of *Queen Elizabeth* is now 797 personnel, a significant increase from the c. 650 originally envisaged, which proved to be insufficient. Including flag staff, air group personnel and various specialists – for example, sailors from the aircraft carrier USS *John C. Stennis* (CVN-74) to prepare ordnance – she has a complement approaching 1,600 for CSG21. A company of Royal Marines from 42 Commando also deployed with CSG21, bringing the number of personnel for the whole force to about 3,700. The carrier strike group represents a significant fraction of the UK's military strength and it is hoped that CSG21 will impress both potential allies and adversaries.

The *Queen Elizabeth*'s younger sister, *Prince of Wales*, commissioned on 10 December 2019 but her progress into service was interrupted when in May and October 2020 she suffered flooding incidents. Subsequent repairs cost £3.3m. After over a year alongside at Portsmouth she returned to sea on 30 April 2021. A revised trials, training and work-up programme should result in her becoming operational – with some limitations – and assigned to NATO duties in 2022. From December 2023 the programmes of the two *Queen Elizabeth* class (QEC) carriers will be structured to ensure that one is either deployed or quickly deployable at as little as five days' notice.

A Royal Marine from 45 Commando pictured in Norway in March 2021 wearing trial equipment as part of the Future Commando Force modernisation. The new equipment includes a lightweight modular ballistic helmet, new camouflage white overalls and a C8 Rifle. The Royal Marine Corps is being re-orientated back to its origins as a seaborne raiding force. *(Crown Copyright 2021)*

LITTORAL STRIKE AND FUTURE COMMANDO FORCE

Since the 1970s the Royal Marines (RM) combat capability has centred on the provision of an elite infantry formation – 3 Commando Brigade RM – which was battle tested in the Falkland Islands, Iraq and Afghanistan wars. In the mid-2000s the RN briefly had sufficient specialist amphibious shipping to land the brigade anywhere in the world – only the US Navy/US Marine Corps could better that. However, cuts were made as large-scale amphibious operations ceased to be a priority and a maximum effort now would be the landing of a Commando Group of 1,800 personnel, based around either 40 or 45 Commando (battalion-strength formations) which alternate in the lead role.

The RN has struggled to justify why the RM rather than the British Army should provide an infantry brigade. This resulted in the development of the Future Commando Force (FCF) concept which re-orientates the Royal Marine Corps to its origins as a seaborne raiding force, with smaller but high readiness combat formations. IR2021 approved this as 'Commando Forces 2030', with £40m allocated to facilitate the changes. The Royal Marines have thus survived, although in a smaller form. Key capabilities it will now provide or support are:

- **Littoral Response Groups:** One stationed in the North Atlantic and Mediterranean Sea, and one (from 2023) in the Indian and Pacific Oceans.
- **Maritime Security, Force Protection and Intervention Operations:** A role assigned to 42 Commando.
- **Special Forces:** The 250-strong Special Boat Service (SBS).
- **Special Purpose Task Group:** This term has often been used for RM detachments, but may now refer to a specific formation established in 2021 to complement the SBS.
- **Guarding the UK's nuclear weapons:** A role assigned to 43 Commando.

Littoral Response Group (North) is synonymous with the existing Littoral Strike Group, which was known as the Amphibious Task Group until 2019. This typically consists of an amphibious transport dock (LPD) and a 'Bay' class auxiliary landing ship. The future LRG (East) will centre on a 'Bay' class vessel with 'littoral strike ship' (LSS) upgrades. Each group will normally embark a 'Strike Company' of c. 150 RM and Army personnel.

In an emergency a QEC carrier can act as a landing platform helicopter (LPH), embarking and landing by helicopters a RM Commando. HMS *Ocean* fulfilled this role until she was sold to Brazil in 2018. It was planned that *Prince of Wales* would receive modifications during her construction to improve her utility as an LPH, but most were cancelled due to cost and a realisation that she was too valuable to be risked in hostile coastal waters.

FORWARD DEPLOYMENT

IR2021 includes the concept of Global Hubs, a network of British military bases across the world to support forward-deployed forces and provide a persistent 'forward presence'. The RN is a key enabler of this vision and will be deployed around the globe to an extent not seen since the 1960s. Measures already implemented include:

- A permanent maritime presence (using a diverse range of RN and RFA ships) in the Caribbean, South Atlantic and Antarctic – utilising

The concept of forward deployment is becoming increasingly important for the Royal Navy. These two images – both dating from 2020 – show the primary casualty receiving ship RFA *Argus* and the 'River' class Batch 2 OPV *Medway* during operations in support of British overseas territories in the Caribbean as part of the Atlantic Patrol Tasking North mission. *(Crown Copyright 2020)*

and protecting British Overseas Territories.

- An offshore patrol vessel (OPV) based at Gibraltar for duties in the western Mediterranean and off the West African coast.
- A frigate and a minehunter squadron operating from the naval base HMS *Jufair*, located in Bahrain, in the Arabian Gulf.
- Establishment of the UK Joint Logistics Support Base at the Indian Ocean port of Al Duqm in Oman, which can berth QEC carriers and nuclear submarines.

Outstanding is the identification of a regional hub and naval facility in the Indo-Pacific region for OPV basing from late 2021, followed by a LSS in 2023. Singapore is a leading candidate due to its geographical location, excellent dockyard capabilities and because the UK already has a small naval facility there (the British Defence Singapore Support Unit), which supports visits by both RN and allied warships.

In 2020 the RN's Fishery Protection Squadron was renamed the Overseas Patrol Squadron, recognising that its five new 'River' class Batch 2 OPVs will rarely be in UK waters. Three are already forward deployed – *Forth* to the Falkland Islands, *Medway* to the Caribbean and *Trent* to Gibraltar – with *Tamar* and *Spey* to head to the Far East soon. The capabilities of the Batch 2 OPVs are being stretched in the forward presence role and it is planned that, by 2030, they will be replaced by five Type 31 general-purpose frigates.

NAVAL CONSTRUCTION PROGRAMME

Aggregating IR2021 with previous announcements, ships under construction or planned in mid-2021 include:

- The last three (of seven) *Astute* class nuclear submarines – all under construction.
- Four *Dreadnought* class ballistic missile submarines – two under construction, two ordered.
- Eight Type 26 anti-submarine frigates – three under construction and five planned.
- Two fast patrol craft for basing in Gibraltar (HMS *Dagger* and *Cutlass*) – under construction.
- Five Type 31 general-purpose frigates – ordered.
- Five Type 32 frigates – planned.
- Three fleet solid support ships (FSS) – planned.
- A multi-role ocean surveillance ship (MROSS) – planned.

Completion of the *Astute* class submarine programme has been beset by delays but progress is now being achieved. This shows the fifth member of the class, *Anson*, being rolled out of the Devonshire Dock Hall in Barrow-in-Furness on 19 April 2021. *(BAE Systems)*

- Conversion of a 'Bay' class landing ship to a littoral strike ship (LSS) – planned.
- A 'National Flagship' – planned.

In addition, concept studies are underway for several other projects:

- SSN(R) nuclear attack submarine – to replace the *Astute* class from 2035 (although 2040 seems more realistic).
- Type 83 destroyer – to replace the Type 45 class from the late 2030s.
- Multi-Role Support Ships (MRSS) – to replace the *Albion* and 'Bay' classes in the early/mid-2030s.

Annual expenditure on warship construction is projected to double to £1.7bn (US$2.4bn) between 2020 (a low point after the completion of several major projects) and 2025; this probably excludes submarines but includes auxiliary ships.

Construction of the *Astute* class submarines has been plagued by problems and delays, and the service life of the old *Trafalgar* class submarines *Talent* and *Triumph* has been extended to preserve a minimum force level of six boats. The commencement of SSN(R) will avoid a significant gap in the workload of BAE's submarine design team post completion of work on the *Dreadnought* class.

HMS *Glasgow*, the first of the three Batch 1 Type 26 'City' class frigates that have already been ordered, should be lowered into the water at BAE Systems Govan shipyard in the coming months but won't enter service until the middle of the decade. An order for five Batch 2 vessels is likely in 2023, with one Type 26 entering service every 15–18 months through to the mid-2030s. It is almost certain that the BAE Systems' Clyde shipyards will

The component sections of the first Type 26 frigate, *Glasgow*, have been joined together at BAE Systems' Govan yard and will be floated out in the coming months. *(BAE Systems)*

then switch to the construction of Type 83 destroyers, thus avoiding a disruptive break in work. It is not known whether the Type 83 will use the same hull as the Type 26, a stretched version, or a completely new hull design.

The decision to progress the Type 83 project to the Assessment Phase makes it unlikely that the six Type 45 class destroyers will receive major mid-life capability upgrades such as adding an anti-ballistic missile (ABM) capability. However, final rectification of previous propulsion plant problems is now underway and it also seems likely that there will be further improvements to their combat and air-defence missile systems.

It was announced by the MoD on 12 September 2019 that a consortium led by Babcock had been selected to the build the Type 31 frigate – now called the 'Inspiration' class. The successful 'Arrow 140' design (139m long and displacing 5,700 tons) is derived from the Danish *Iver Huitfeldt* design. A contract for five frigates was awarded on 15 November 2019 and Babcock should begin fabrication of the first, *Active*, in late 2021 at a new construction shed built at its Rosyth shipyard. Under the current optimistic schedule, she will be launched in 2023 and enter service in 2027, potentially beating *Glasgow* as the first new RN frigate since 2002! All five should be operational by 2030.

A surprise announcement associated with IR2021 was the Type 32 frigate; 'designed to protect territorial waters, provide persistent presence overseas and support our Littoral Response Groups'. Hints that the class will be built at Rosyth combined with reports that the MoD is studying a 'platform for autonomous systems' have resulted in speculation that the Type 32 will be a variant of the Type 31 design adapted as a 'mothership' for autonomous

An updated computer graphic showing the nearly-finalised Type 31 frigate design. A comparison with the image published in *WNR*2020 reveals many interesting changes, including a reduction from twenty-four to just twelve Sea Ceptor VLS cells and the replacement of two light cannon forward of the bridge with Bofors Mk 4 40mm gun systems in the 'B' and 'Y' positions. No anti-ship missiles are fitted. *(Babcock International)*

underwater, surface and airborne vehicles. In particular, it might operate the automated Mine Hunting Capability (MHC) that has been under development since 2010 as part of the UK-French Maritime Mine Counter Measure (MMCM) programme. Three operational MHC sets – including 12m unmanned surface vehicles – were ordered by the UK in 2020 at a cost of £184m and the first will be delivered in late 2022. It will then undergo operational evaluations before entering service. However, if the size of the RN's escort force is to increase to twenty-four by the early 2030s – as Prime Minister Johnson claims will happen – the RN will need to operate the Type 32s as traditional frigates. Also, larger vessels similar to the current 'Bay' class might be more effective and offer better value as motherships for supporting MHC and other maritime autonomous systems.

The FSS project dates to the early 2000s, when it was part of the ambitious Military Afloat Reach and Sustainability (MARS) programme. The ships will replace the supply ship RFA *Fort Victoria* and two ships that have recently been retired, namely *Fort Rosalie* and *Fort Austin.* Whilst the project is critical to Carrier Strike, a lack of funding has repeatedly delayed progress. When in 2020 it became clear that the FSS order would go to an overseas shipyard as no British shipyard could compete on cost, politics intervened and no contract was awarded. The procurement process resumed in May 2021 using a slightly less demanding specification – in particular the reduction of a Heavy Replenishment at Sea (HRAS) capability but adding a requirement that a 'significant proportion' of the c. 40,000-ton ships must be built in the UK. A manufacturing contract is expected to be made 'within two years', a relaxed timescale for an already long-delayed programme.

It was announced in February 2019 that the RN would receive two new littoral strike ships. IR2021 has reduced this plan and allocates £50m for a limited conversion of one 'Bay' class landing ship to the LSS role – probably RFA *Cardigan Bay.* She will be fitted with a permanent hangar and better aviation facilities for embarked helicopters and UAVs; new command, control and communication capabilities; improved medical facilities; and enhanced accommodation and storage for the embarked military force.

The MRSS Project entered the Assessment Phase in 2020. Up to six ships will provide 'platforms to deliver Littoral Strike, including Maritime Special Operations', replacing two *Albion* and three 'Bay' class ships that are all due to leave service between 2031 and 2034. MRSS will be a relatively large ship intended for global operations, with internal vehicle and stowage decks, extensive accommodation, and good aviation capabilities. Although optimised for amphibious warfare, the likely provision of secondary logistic and replenishment at sea capabilities will add versatility.

In September 2019 BMT announced the ELLIDA concept, a family of multi-role logistics vessels. Shown is the ELLIDA 200 design – displacing around 23,000–24,000 tonnes with a length of 200m and beam of 30.5m. A variant is a strong contender for the Multi-Role Support Ship (MRSS) programme, which it has undoubtedly influenced. *(BMT)*

MROSS is an urgent operational requirement with an unusually rapid procurement process; a MoD press release states that 'the ship, which will come into service by 2024, will be a surface vessel, with a crew of around 15 people'. Whilst its primary role will be to protect the UK's undersea infrastructure using advanced sensors and remotely operated/autonomous undersea drones, the ship will also be useful in gathering intelligence about potential threats. It is unclear whether the ship will be a new build or a conversion but a number of British shipyards such as Cammell Laird and the revitalised Harland & Wolff's Appledore facility are believed to be interested in the work. 1SL has implied that a second ship might be purchased.

IR2021 includes no direct replacement for the primary casualty receiving ship (PCRS) RFA *Argus*, which recently completed a refit to extend her service life to 2024. Another notable omission is a replacement for the RN's only large ocean survey vessel, *Scott*, which is due to leave service in 2022. The 'Admiralty Charts' that have been sold by the United Kingdom Hydrographic Office for over 200 years – albeit now usually in electronic rather than printed form – and which are still used by most of the world's ships, will thereafter have fewer updates based on source data provided by the RN.

In 1997 the Royal Yacht *Britannia* was controversially decommissioned without replacement. On 29 May 2021 the Prime Minister announced plans to order a 'National Flagship' manned by the RN. Whilst primarily intended to 'showcase' and promote Britain, it will potentially host the Royal family for overseas visits.

FLEET AIR ARM (FAA)

IR2021 includes only a few decisions affecting the FAA, the most notable being the withdrawal from service of the Hawk T 1 jet trainer. This implies that 736 Naval Air Squadron (NAS) will decommission in the next few years. The primary role of the squadron is to simulate aircraft and missile attacks on ships during training; presumably other aircraft and drones will be used instead.

The 2030 out-of-service date of the FAA's Merlin HM 2 helicopters is rapidly approaching, although this seems likely to be extended. IR2021's announcement that a new medium-size helicopter will be procured for the RAF and Army is significant, as a naval version will be a candidate for the replacement of Merlin.

April 2023 is the target date for 809 NAS commissioning as the UK's second operational F-35B Lightning squadron, 13 years after the FAA last operated front-line fixed-wing aircraft. The carrier strike programme originally envisaged four twelve-aircraft front-line squadrons being formed from a purchase of 138 aircraft. IR2021 does not state how many F-35Bs the UK will buy above the forty-eight currently due for delivery by the end of 2025, but senior officials suggest that a final total of sixty–eighty aircraft is likely – so fewer and/or smaller squadrons seem inevitable.

May 2021 saw the helicopter-borne Crowsnest surveillance and command system being deployed as part of CSG21, despite not achieving its initial operational capability (IOC) milestone due to software-related problems. Crowsnest is only a modest upgrade of the airborne early warning (AEW) system fitted to the previous Sea King ASaC 7, with a mechanically-scanning radar – Searchwater 2000 – that dates to the 1970s. This is recognised by an expected short service life; until just 2029.

The FAA has a vision – *Future Maritime Aviation Force 2030* – which seeks greatly improved capabilities.[10] This cryptically envisions that 'In 2030, medium organic UAS [unmanned air systems] will deliver most FIND in the Littoral and Maritime,

A F-35B Lightning II strike fighter from the Royal Air Force's 617 Squadron pictured on the flight deck of the aircraft carrier *Queen Elizabeth* during night-time operations in the course of the CSG21 deployment. By 2023, 809 Naval Air Squadron (NAS), the second operational F-35B squadron, is also expected to be operational. *(Crown Copyright 2021)*

teamed with crewed air for FIX/STRIKE, and complemented by Maritime Protector. ASW Merlin will regain mass with the transfer of AEW capability to an alternate platform. Together with enhanced FW in the Carrier Air Wing, Maritime Aviation will be reconstituted to match evolving peer threats.'

Project Vixen is exploring how fixed-wing unmanned air systems (FWUASs) might be incorporated as part of a QEC carrier air group by 2030. FWUASs could be used for missions including surveillance, air-to-air refuelling, electronic warfare and strike. It is likely that Vixen will be based on the Mosquito UAS being developed as part of the RAF-led Lightweight Affordable Novel Combat Aircraft (LANCA) programme.

Mosquito and other FWUAVs able to fill the Vixen role will be relatively big and heavy. As a result, in February 2021 the MoD issued a Request For Information (RFI) to assess the 'availability of electromagnetic catapult and arrestor wire systems to launch aircraft' within 3-5 years. The RFI defines a maximum launch weight of 25 tonnes and maximum landing weight of 21 tonnes. It is possible that *Prince of Wales* will be fitted with a revised flight deck configuration incorporating an electromagnetic catapult and arrestor wires during her first major refit in the late 2020s. These will be solely for FWUAS; F-35B Lightning IIs will continue to use the bow ski-jump.

A separate rotary-wing Project Proteus is urgently examining the use of rotary-wing unmanned air systems (RWUASs) for anti-submarine and littoral strike missions – supplementing or partially replacing the capabilities of the Merlin HM 2 and Wildcat HMA 2 helicopters. The first RWUASs could enter service as early as 2023.

Prince of Wales (foreground) and *Queen Elizabeth* sail in company for the first time prior to the latter's deployment on CSG21. It is possible that *Prince of Wales* will be fitted with a revised flight deck configuration incorporating an electromagnetic catapult and arrestor wires for UAV operations during her first major refit in the late 2020s. *(Crown Copyright 2021)*

A variety of unmanned and autonomous vessels are being trialled by the Royal Navy. This image shows MADFOX (Maritime Demonstrator For Operational eXperimentation), which was delivered in March 2021. *(Crown Copyright 2021)*

The new Thales Martlet missile has carried out initial test firings from the Leonardo Helicopters Wildcat helicopter and has been deployed on CSG21. *(Thales)*

NELSON AND NEMISIS

NELSON (software focused) and NEMISIS (autonomy and experimentation focused) are programmes to introduce innovative and transformational technologies to the RN. A range of projects have been launched – with titles such as NavyX and DARE – tasked with delivering capabilities ranging from artificial intelligence to the 3D printing of spare parts.

The RN's ambitions regarding autonomous underwater, surface and airborne systems have been constantly hampered by a lack of funding. IR2021 promises to back initiatives such as the RN's autonomous submarine MANTA extra-large unmanned underwater vehicle (XLUUV) technology demonstrator, which was delivered by MSubs Ltd in February 2021. Just a month later the NavyX team accepted the autonomous vessel MADFOX (Maritime Demonstrator For Operational eXperimentation), the latest in a series of autonomous surface vessels. These vessels are now being tested across a range of military operations, from surveillance to force protection.

The RN has also established Maritime Enterprise Planning Groups to work with UK industry on advanced concepts. For example, 'Atlantis' envisages a semi-autonomous mothership submarine operating from c. 2040 in tandem with remote, autonomous and un-crewed platforms.

WEAPONS

IR2021 confirmed that:

- A short-term replacement for the Harpoon anti-ship missile – the Interim Surface to Surface Guided Weapon (I-SSGW) – will be ordered.[11]
- Work on a replacement for the Stingray lightweight torpedo has begun.
- Additional Sea Viper air-defence missiles (aka the MBDA manufactured Aster) will be ordered – it has long been suspected that the RN's stock is inadequate. The latest Aster 30 Block 1NT has an ABM capability but Type 45 destroyers cannot utilise this without upgrades.

Conversely, IR2021 did not announce an anticipated order of cruise missiles for the Type 26 'City' class but this is still likely.

The RN is purchasing several new gun systems. As well as the 5in 62-calibre Mk 45 naval gun being fitted to the Type 26 class, BAE Systems has received orders to manufacture Bofors 40 Mk 4 and Bofors 57 Mk 3 guns for the Type 31 frigates.

Work continues on the Dragonfire laser directed energy weapon system which has been under development since 2017. The delayed sea trials of a demonstrator system are now planned for 2023.

The target in-service dates of the (light) Martlet and (heavy) Sea Venom anti-ship missiles to be carried by Wildcat helicopters have also repeatedly suffered from delays. Whilst not formally announced, Martlet was approved in early 2021 for operational use by Wildcat helicopters taking part in CSG21, leading to intensive last-minute training. The latest in-service date for Sea Venom is 2022.

Finally, after 41 years of sterling service (including in the 1982 Falklands War) the veteran Sea Wolf missile system was retired in 2020, replaced by Sea Ceptor.

PERSONNEL AND TRAINING

Since 2015 the target trained strength for the Royal Navy and Royal Marines has barely changed. As of 1 April 2021 this was 30,760 regulars compared to an actual strength of 29,460 – a shortfall of 4.2 percent. However, the trend is positive, as the deficit was 5.2 percent a year earlier.

After many years of disappointing RN recruitment (despite spending over £40m p.a. on advertising), a significant upsurge began in 2019. By early 2020 the primary shore-based training establishment (aka 'stone frigate'), *Raleigh*, was reaching its intake capacity of sixty new-entry ratings a week. Rather than turn away enthusiastic applicants, another establishment (*Collingwood*) began training new entries for the first time since the Second World War, and extraordinarily the *Britannia* Royal Naval College – normally reserved solely for initial officer training – also trained 100 ratings in 2020. As a result, the total strength – trained and untrained – of the RN/RM surged from 32,760 on 1 April 2020 to 33,850 on 1 April 2021 – an increase of over 1,000.

Improved recruitment smoothed the award in December 2020 of a 12-year £2bn contract for training RN personnel to a consortium led by Capita. This marks a much-needed shake-up after decades of under-investment in training and will also free uniformed personnel for sea rather than shore billets. But an emerging problem is the reducing number of minor RN warships available for newly-promoted officers and senior rates to gain command or charge experience. Other navies have encountered the pitfalls of appointing inexperienced personnel directly to senior appointments on major warships.

IR2021 apparently continued a trend of reducing the proportion of Royal Marines in favour of more sailors. The actual number of RMs fell from 7,110 in 2015 to 6,790 in 2017. No official numbers have been since published since, but a strength of about 6,000 RMs is likely by 2022.

Whilst not well regarded by some in the RN, the c. 4,000 strong Royal Naval Reserve continues to survive. It is cheap to maintain and has a relatively high profile in fifteen towns and cities around the UK, most of which would otherwise have no naval presence. RNR establishments are being provided with modern equipment to help make reservists more effective more quickly when supplementing the crew of operational warships.

A FINAL COMMENT

A lack of warships and submarines makes it impossible for the Royal Navy to fully meet the UK government's 'Global Britain' objectives in the short-term. Whilst the CSG21 deployment is impressive, it has absorbed much of the RN operational strength in 2021 to the detriment of other more mundane tasks. It will also take two years for the RN to regenerate a similar capability.

It will not be until around 2030 that the RN will start reaping the rewards of the naval construction programme described in IR2021, even assuming that it is fully implemented. Moreover, the United Kingdom continues to suffer from serious internal political tensions which could even lead to the fragmentation of the country in the next few years. In that event, it is impossible to forecast how the Royal Navy will need to change and adapt.

Notes

1. Her Majesty's Naval Service consists of the Royal Navy, Royal Marines, Royal Fleet Auxiliary, Royal Naval Reserve, Royal Marines Reserve and the Naval Careers Service. This article uses the term 'Royal Navy' as a commonly accepted title for all of these, unless a branch is explicitly named.

2. See 'PM Statement to the House on the Integrated Review: 19 November 2020' available online by searching www.gov.uk.

3. *Global Britain in a competitive age: The Integrated Review of Security, Defence, Development and Foreign Policy, Cm. 403* (London: HMSO, 2021). This can be accessed by searching: www.gov.uk

4. *Defence in a Competitive Age. Cm. 411* (London: HMSO, March 2021). Accessible at: www.gov.uk/government/publications/defence-in-a-competitive-age

5. IR2021 should be read in conjunction with the *Integrated Operating Concept* (Bristol: MoD, 2020) and the *UK Joint Doctrine Publications*, which are currently being updated. Some are available by searching online.

6. Admiral Radakin made a major speech to the First Sea Lord Sea Power Conference on 19 May 2021. See www.gov.uk/government/speeches/first-sea-lord-sea-power-conference-speech

7. See the author's article 'Royal Navy: A Thin Blue Line' in *Seaforth World Naval Review 2016* (Barnsley: Seaforth Publishing, 2015), pp.74–87.

8. The MoD is required to provide Parliament with an annual progress report on the *Dreadnought* programme; this is essential reading for analysts. The 2020 update can be found at: commonslibrary.parliament.uk/research-briefings/cbp-8010/

9. A summary of the status of the whole carrier strike programme can be found in the report *Carrier Strike – Preparing for deployment* (London: NAO, June 2020) available at: www.nao.org.uk/report/carrier-strike-preparing-for-deployment

10. An informative presentation by the RN/FAA Maritime Enterprise Planning Group entitled *Navy Develop Carrier Strike and Maritime Aviation* entered the public domain in March 2021.

11. On 8 March 2019 the Ministry of Defence published a requirement for a 'Next Generation Surface Ship Guided Weapon (SSGW)' to equip Royal Navy vessels, with first delivery of missiles required by December 2023. After delays, a formal request for tenders is expected in mid-2021.

52

3.1 SIGNIFICANT SHIPS

BOUCHARD CLASS OPVS

French Technology modernises the Argentine Navy

Author:
Bruno Huriet

In May 2020, a Naval Group-built offshore patrol vessel (OPV) that had only recently been commissioned into the Argentine Navy intercepted and arrested a foreign jigger illegally fishing in the country's exclusive economic zone (EEZ). The ship – ARA *Bouchard* – is the first to be delivered under a four-vessel contract signed between the Argentinian government and the French shipbuilder in 2018. The agreement encompassed one OPV – *L'Adroit* – that was already afloat and then being operated by the French Navy and which was to be recommissioned as ARA *Bouchard*. The three other ships were to be newly built. The ships are the smallest units within Naval Group's extended 'Gowind' family, *L'Adroit* being first of the type.

CLASS ORIGINS

The name 'Gowind' appeared in the news for the first time in 2004 at the Euronaval exhibition in Paris. At that time, Naval Group (then DCNS) had gained success in exporting frigates and submarines but lacked a portfolio of smaller ships.[1] Accordingly, a range of corvettes – essentially well-armed 'mini-frigates' in the 1,000–3,000-tonne size range – was developed to fill this gap. Several proposals for these ships were subsequently discussed with a range of countries and Bulgaria even signed a letter of intent to order a variant. However, none of these discussions ultimately reached a firm conclusion.

Left: One of four 'Gowind' or *Bouchard* class OPVs ordered by Argentina from France's Naval Group, *Piedrabuena* is seen approaching the final stages of fitting out at the Piriou yard at Concarneau in December 2020. She is the first new-built warship to be delivered to the Argentine Navy for many years. *(Bruno Huriet)*

Accordingly, at the 2008 Euronaval exhibition, Naval Group presented a modified 'Gowind' family offering an enlarged range of ships. In contrast with the original concepts, there was greater focus on sovereignty enforcement missions than actual combat. Key design drivers included straightforward, economic operation and maintenance, the possibility of local construction through technological transfer and, of course, a competitive price. The 'low end' of the revised family was the 'Gowind Control', a lightly-armed, c. 1,000-tonne OPV optimised for constabulary operations. Further concept designs of increasing size and sophistication culminated in the 'Gowind Combat', a 2,500-tonne

A Naval Group computer-generated image of a 'Gowind corvette. The 'Gowind' concept was initially focused on a range of well-armed mini-frigates but was later expanded to encompass a number of constabulary OPV designs. *(Naval Group)*

Assembled at Lorient in Brittany under Project Hermes, the first 'Gowind' OPV – *L'Adroit* – was loaned to the French Navy on completion. This picture was taken in November 2011, shortly after she had entered *Marine Nationale* service. *(Naval Group)*

corvette based on the original concepts and incorporating a full combat suite of guns, missiles and sensors.

However, the 'Gowind' family remained paper ships, with no firm orders being placed. The French Navy – usually Naval Group's first and principal customer – was heavily committed to the FREMM frigates of the *Aquitaine* class. It was therefore not in a position to place orders for new OPVs in the face of severe budget cuts, in spite of a pressing requirement for ships of that type. The lack of a reference order from such an important customer, in turn, impacted prospects for overseas contracts.

The limited prospects for French Navy orders and the consequent increased importance of export business led Naval Group's newly-appointed CEO, Patrick Boissier, to support an innovative approach. In order to offer potential customers a 'sea proven' ship, he announced, early in 2010, the launch of Project Hermes. Under this, a 'Gowind' OPV would be built with Naval Group's own funds and then made available on loan to the French Navy to allow them to evaluate the ship and its systems. The agreement suited both parties. The navy was provided with the opportunity to fill part of the gap caused by the retirement of older ships without replacement with a free, modern OPV that would allow it to explore the potential offered by the latest technological innovations. Meanwhile, Naval Group would be able to place the 'sea proven' label on the 'Gowind' family whilst benefiting from the publicity gained by the new ship's visits to potential customer countries and naval exhibitions.

Once this decision was made, the OPV – named *L'Adroit* – was built very quickly at the group's shipyard at Lorient in Brittany with the support of a dedicated Naval Group project team. The ship was handed over to the French Navy after the completion of builder's trials on 21 October 2011, less than 18 months after first steel-cutting on 7 May 2010. Although an initial loan period of three years was envisaged, the term was subsequently extended on a number of occasions, with *L'Adroit* not finally being returned until 2018.

Meanwhile, in 2013, Naval Group had entered

into the Kership joint venture with commercially-focused shipbuilder Piriou to strengthen its ability to supply vessels focused on constabulary and other non-combat related government roles. This development resulted in a decision to transfer the Naval Group OPV portfolio into the new alliance and reserve the 'Gowind' branding for the combat mission-focused corvettes to which it had first been applied. Accordingly, the *L'Adroit* OPV 87/OPV 90 design now forms the largest of a series of patrol vessels ranging from 45m to 90m in length that are marketed by the Kership organisation.[2]

L'Adroit pictured at Toulon in April 2018. The ship's period on loan to the *Marine Nationale* was extended on a number of occasions and it was not until 2018 that she returned to Naval Group control. *(Bernard Prézelin)*

THE ARGENTINIAN CONTRACT

Argentina has a coastline of nearly 5,000km and an extensive EEZ of over a million square kilometres, even excluding ongoing claims to territories currently under British administration in the South Atlantic. These waters are rich in fish such as hake and squid but suffer from severe illegal, unreported and unregulated (IUU) fishing, affecting Argentina's own fishing fleet and endangering stocks. With Argentina's sovereign rights for the purposes of exploring, exploiting, conserving and managing fisheries resources within its EEZ firmly established under the current United Nations Convention on the Law of Seas (UNCLOS), protection of these and other resources is clearly an important maritime task. In Argentina, responsibility for policing the EEZ is shared between the coastguard – the *Prefectura Naval Argentina* – and the country's navy – the *Armada de la República Argentina* or ARA. The need for dedicated vessels optimised for lengthy patrols across the vast expanses of these waters without the need for resupply has long been recognised.

Given this situation, the navy hoped to launch a new programme for OPV construction as early as the mid-2000s. Initially referred to locally as *patrulleros de altamar* (PAM) or high seas patrol vessels and later *patrulleros oceánicos multipropósito* (POM) or multi-purpose oceanic patrol vessels, reports at the time suggested that between four or five ships would be built locally to the German company Fassmer's OPV80 design.[3] In the event, this programme did not proceed and the navy acquired four second-hand offshore support vessels from Russia in 2014 to meet its most urgent patrol and supply mission requirements.

However, the search for a new OPV design continued. In 2015, negotiations took place as part of a wider trade agreement with China for three vessels similar to the Nigerian P18N *Centenary* class, itself a variant of the People's Liberation Army Navy's Type 056 corvette. Further talks were conducted with a range of suppliers against a backdrop of significant pressure from illegal activity in the country's fishing grounds. Ultimately, an understanding was reached with France that envisaged the acquisition of four 'Gowind'-type OPVs. According to contemporary reports, two were to be built in France and two assembled in Argentina by the local Tandanor shipyard under a transfer of technology agreement.

Nearly two years were to elapse until the signature of a finalised agreement, largely due to complex negotiations relating to the programme's financing. A contract was ultimately signed with Naval Group on 22 June 2018 after further high-level discussions. *L'Adroit* – the ship first delivered to the French Navy in 2011 – would be transferred after a dry docking and overhaul. Three additional, enhanced ships were to be built in France by the Kership joint venture at yards in Brittany. Overall contract value amounted to c. €300m (c. US$360m). Following completion of financing discussions, Naval Group announced that the contract had taken formal effect on 13 February 2019.

DESIGN PARTICULARS

The *Bouchard* class OPV 87 is essentially designed to Classification Society standards. The ships comply with International Maritime Organization (IMO) rules with respect to stability and are built according to Bureau Veritas rules under the classification society's surveillance.[4] Both hull and superstructure are fabricated from steel, with other key design information summarised in Table 3.1.1.

Platform: As for all ships of its type, the *Bouchard* class is optimised to ensure the efficient and cost-effective surveillance and protection of its country's EEZ, performing tasks ranging from fishery protection and border control through to counter-trafficking, pollution control and search and rescue operations. The platform design was therefore drawn up with these requirements in mind.

L'Adroit (now ARA *Bouchard*) pictured in rough seas in the course of French Navy service in 2017. An ability to maintain excellent seakeeping performance and stability in adverse weather conditions is an important characteristic of the *Bouchard* class. *(François Bogaert/ Marine Nationale)*

The hull of the third *Bouchard* class OPV, *Storni,* pictured in the course of construction. The ships utilise an all-steel hull whilst the three vessels specifically ordered for the Argentine Navy also incorporate reinforced hulls to provide a limited ice-going capacity. *(Bruno Huriet)*

This 2020 view of *Piedrabuena* at Concarneau demonstrates the essentials of the *Bouchard* class design. A prominent design feature is the location of the bridge high atop a relatively short superstructure, providing 360° panoramic vision. The bow thruster and fin stabilisers were an Argentine-requested addition to the original design and are not fitted to *Bouchard*, which is, however, equipped with a FLUME passive roll damping system. *(Bruno Huriet)*

A major design consideration was an emphasis on excellent seakeeping characteristics and stability. On *L'Adroit/Bouchard*, a FLUME passive roll damping system was installed. This uses connected lateral tanks, with the shift in water dampening rolling. However, the three new ships built for the specific Argentine order adopted a more traditional arrangement based on underwater fin-type stabilisers. These three ships have also been built with reinforced hulls to provide a limited ice-going capacity.

A prominent design feature is the location of the bridge high atop a relatively short superstructure. This arrangement provides 360° panoramic vision whilst releasing space for the proportionately large helicopter deck and the rigid-hulled inflatable boat (RHIB) launching ramps aft. In order to optimise crew efficiency, the combat information centre (CIC) and bridge are co-located, thereby providing the bridge team with ready access to

The *Bouchard* class is fitted with a proportionately large helicopter flight deck and twin launch and recovery ramps for embarked RHIB type interceptors. The system can operate in conditions up to Sea State 4 at speeds of between 6 and 12 knots, with time for launch or recovery less than five minutes. *(Naval Group)*

the tactical picture collated by all the ship's sensors.

The hull is of steel construction and measures 87m overall (81.5m between perpendiculars). It is divided into nine main watertight zones. There are five principal decks, viz.

- **10 Deck:** This is the lowest deck level. Much of the forward part houses crew berthing quarters, including two to ten bunk areas and wet spaces. The engine room is amidships and compartments for stores further aft.
- **00 Deck (Main Deck):** Moving from forward to aft in the ship, this incorporates further stowage compartments followed by cabins for the ship's officers and senior ratings and then the galleys and mess rooms. The upper level of the engine room and the machinery control room (MCR) is followed by offices, additional accommodation spaces and a changing room for divers. The RHIB launch and recovery system is furthest aft.
- **01 Deck (Weather Deck):** This carries the main

Table 3.3.1.

BOUCHARD CLASS OPV PRINCIPAL PARTICULARS

Building Information (*L'Adroit*/*Bouchard*):

Fabrication Commenced:	7 May 2010 **Launched:** 18 May 2011[1] **Delivered:** 21 October 2011[2]
Builders:	DCNS (Now Naval Group) at its Lorient yard[3]

Dimensions:

Displacement:	1,650 tons full load displacement.
Overall Hull Dimensions:	87.0m x 13.6m x 3.4m. Length at waterline is 81.7m.

Equipment:

Armament:	1 x Leonardo 30mm Marlin WS. 2 x Reutech 12.7mm Sea Rogue.
Aircraft:	1 x 5-ton helicopter (platform and hangar). Flight deck can handle helicopters of up to 10 tons in weight.
Principal Sensors:	Terma Scanter 6002 surveillance radar. Navigation radar. Safran Vigy Engage electro-optical fire-control system.
Combat System:	Naval Group Polaris combat management system. Integrated communications suite.
Other:	2 x 9m RHIBs for boarding operations. 1 x 4m rescue RHIB. Flight deck can transport up to two containers with various equipment.

Propulsion Systems:

Machinery:	CODAD. 2 x ABC 12 DZC diesels each rated at 2.7MW. Two shafts. (*Bouchard*) CODAD. 2 x ABC 16 DZC diesels each rated at 3.5MW. Two shafts. Bow thruster. (Other class members)
Speed:	Maximum continuous speed is c. 20 knots. This is achievable at 80% of MCR in the three later vessels. Range is 7,500 nautical miles at 12 knots.

Other Details:

Complement:	Normal crew in Argentine Navy service is c. 32–40. Accommodation is provided for 59 personnel.

BOUCHARD CLASS LIST

NAME	PENNANT	FIRST STEEL CUT	LAUNCHED	DELIVERED	BUILDERS
Bouchard	P51	7 May 2010	18 May 2011	21 October 2011[4]	DCNS Lorient
Piedrabuena	P52	February 2019	1 October 2020	13 April 2021	Piriou Concarneau
Storni	P53	March 2019	10 May 2021	[October 2021]	Piriou Lanester & Concarneau
Bartolome Cordero	P54	October 2019	[September 2021]	[April 2022]	Piriou Lanester & Concarneau

Notes:

1. Refers to formal christening. Physical launch operations were performed during the preceding days.
2. Delivery on loan to the French Navy as *L'Adroit*. The refurbished ship was subsequently transferred to the Argentine Navy on 6 December 2019.
3. The forward part of the hull was fabricated by Piriou at its Concarneau yard. Please refer to the text for the construction of subsequent vessels.
4. Delivery on loan to the French Navy. Formal delivery to the Argentine Navy was on 6 December 2019.

Piedrabuena (2021): Profile and Plan

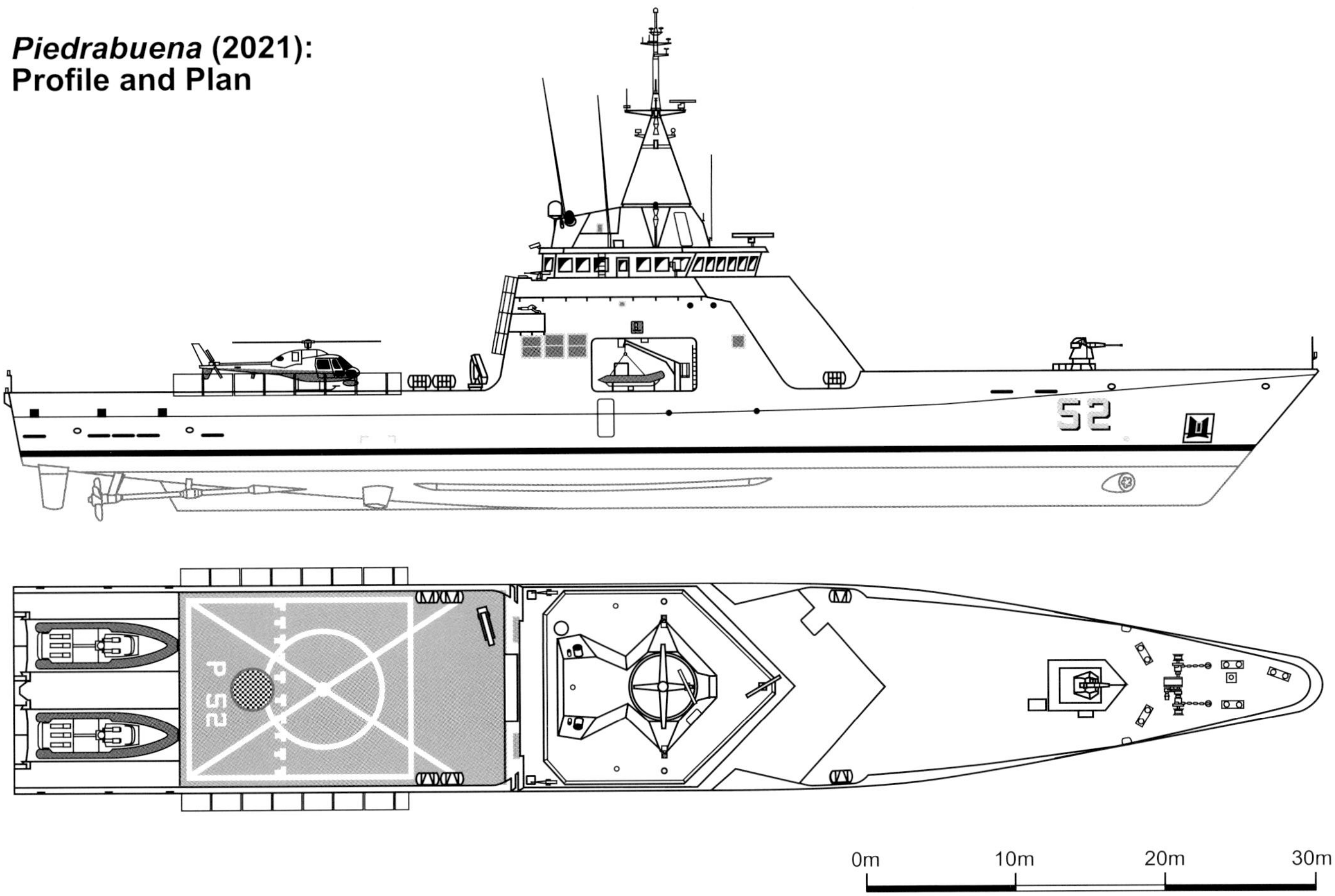

Piedrabuena: Internal Layout

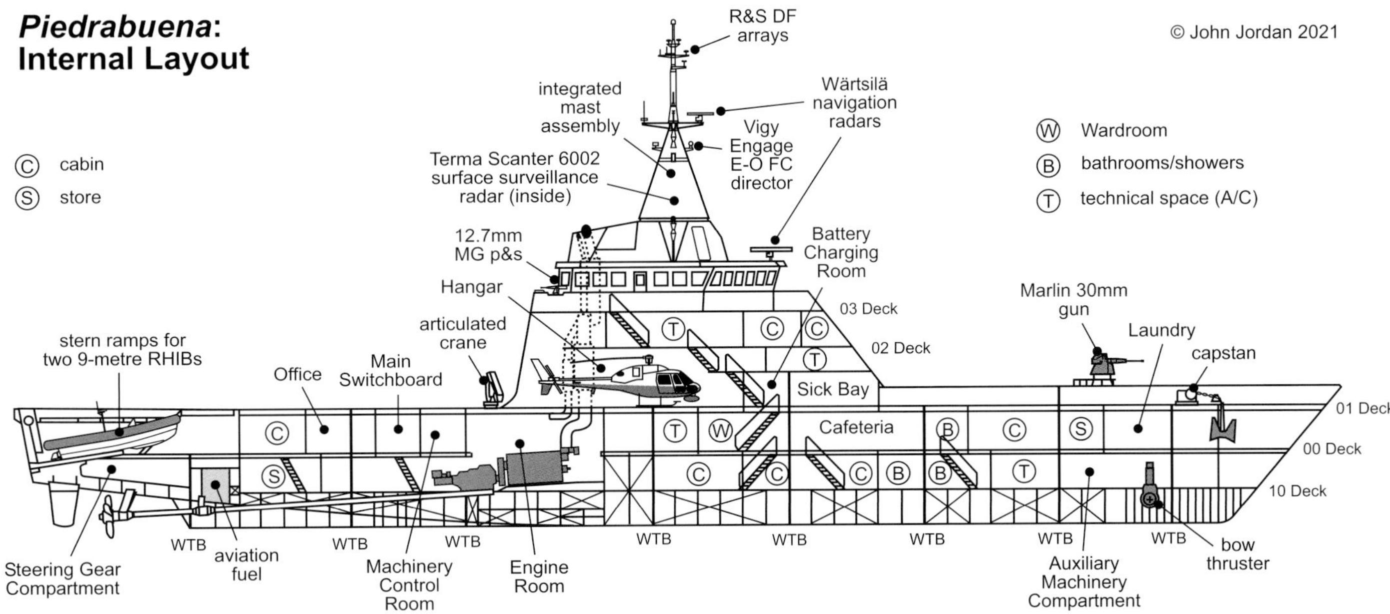

Piedrabuena pictured in the course of preliminary sea trials in March 2021. She is designed around five principal deck levels, with her single damage control zone being sub-divided into two main vertical areas. *(Marc Ottini)*

gun forward of the superstructure, whose base houses an emergency generator and the helicopter hangar, which opens out onto the flight deck. A recess to starboard accommodates a 4m rescue RHIB whilst there is a replenishment at sea (RAS) station with padeyes, a winch and pipe connection points to port. Further aft is the upper point of the RHIB launch and recovery system.

- **02 Deck:** This houses the captain's and executive officer's quarters.
- **03 Deck (Bridge Deck):** This is occupied by the large (120m^2) combined bridge and CIC. Above, a technical room is located at the base of the enclosed mast.

The ship has just one damage control zone, subdivided into two main vertical areas. Monitoring and damage control functions can be performed either from the MCR or the bridge. Automatic fire detection is installed throughout the ship, combining smoke, flame, and temperature detectors. Fire-fighting facilities include a fire main fed by two pumps and additional fixed systems for the protection of specific, more vulnerable compartments. Notably, the magazine can be doused with water sprayers whilst the engine room and radar technical room are equipped with a 'clean agent' fire suppression system using Novec 1230 fire protection fluid.

The 280m^2 helicopter deck is fitted with a landing grid and deck securing fittings, allowing operation of 10-tonne class helicopters such as the NH Industries NH90. However, the hangar – or, more accurately, shelter – can only accommodate a 5-tonne class machine such as the Eurocopter AS 555 Fennec 2 used by Argentina's Naval Aviation.[5] Due to the hangar's limited size, it is not possible to perform maintenance onboard. A 'Mantis' electrical tug supports helicopter handling. Visual landing aids providing a horizontal reference and glide slope are provided to allow both day and night operation. All flight-deck operations are controlled from a dedicated console located in the aft part of the bridge, thereby offering an unobstructed view of the deck.

A picture of *Piedrabuena*'s flight deck and hangar taken at the time of her commissioning on 13 April 2021. The flight deck is able to operate 10-tonne class medium helicopters but the hangar is only able to stow light helicopter types. Evident to the left of this view is an articulated crane and the 'Mantis' electrical tug that supports helicopter handling. *(Naval Group)*

The *Bouchard* class OPVs are powered by a two shaft/two engine CODAD arrangement that utilises Anglo Belgian Corporation DZC diesels. The prototype *L'Adroit/Bouchard* is equipped with the 12-cylinder V 12DZC (left) that can provide up to 2.7MW in continuous mode whilst the other ships have the 16-cylinder 16DZC (right) which can generate 3.5MW. This allows these later ships to achieve maximum designed speed without relying on the diesel's maximum power output. *(Anglo Belgian Corporation)*

Refuelling equipment for Jet A1 aviation fuel is installed. This encompasses an 11m^3 tank; pumping, treatment and distribution equipment; as well as a nitrogen inert gas generator for tank protection.

The class's stern area is dedicated to the launch and recovery systems for two 9m RHIBs. The rapid deployment of inflatable boats is one of the class's most frequent operational requirements, being undertaken to transfer boarding teams, perform rescue at sea, evacuate civilians or support the transfer of Special Forces in as wide a range of sea states as possible. The twin stern ramps and associated handling system have been engineered by BOPP to provide a quicker and safer option to traditional davits. Launch is typically performed whilst the OPV is still underway, with the fully-manned RHIB being pushed by a beam moving on overhead rails and with its hull sliding on low-friction pads through the relevant open stern doors. Once afloat it is towed by the OPV until, with its engine running, it disconnects from the mother ship. When returning, the RHIB's crew catch a towing cable which is deployed by the moving OPV to float astern; once connected the RHIB is winched back onboard. The system can operate in conditions up to Sea State 4 at speeds of between 6 and 12 knots, with time for launch or recovery less than five minutes. The *Bouchard* class are each equipped with two 9m Zodiac 'Hurricane' RHIBs which carry inboard diesel engines.

Engineering Aspects: All the main machinery is located in a single, spacious engine room, the reduced need for survivability allowing a less cluttered layout than the compartmentalised arrangement typical of front-line warships.

The propulsion plant selected has been driven by the need to patrol the high seas in the most economical way possible and is based on a two shaft/two engine CODAD arrangement. Each Anglo Belgian Corporation (ABC) 16 DZC series engine – rated at 3,500kW at 1000 rpm – is connected via a Reintjes reduction gearbox to a Schottel shaft line and a controllable-pitch propeller.[6] In addition to providing a degree of redundancy, the propulsion arrangement also allows a degree of fuel economy that is especially useful on long-range patrol operations, with 12 knots being achievable from one shaft operation. Maximum sustained speed on both shafts is in the region of 20 knots.

The three ships ordered specifically for the Argentine Navy incorporate a bow thruster to improve lower speed manoeuvrability. Shaft-driven generators have also been specified to help meet the increased electrical generating capacity required by this arrangement.

The engine room is also home to a range of auxiliary machinery which includes the two Volvo Penta diesel generators that supply the ship's electrical distribution system, a heating ventilation and air-conditioning (HVAC) water chilling plant, two fresh water reverse osmosis generators and grey and black water storage and treatment plants. The additional emergency diesel generator installed in a compartment on 01 Deck is fitted with battery starting and is designed to come on line automatically in the event of an electrical power failure.

The machinery is controlled and monitored from consoles in the machinery control room located on the main deck. A switchboard room is located just aft. The machinery can also be operated from a dedicated console fitted in the ship's bridge.

Sensors and Bridge Equipment: A Naval Group Polaris combat management system forms the core of the *Bouchard* class's operational capabilities. It essentially gathers and processes the information obtained by each of the ship's sensors, providing the combined bridge and CIC team with a complete picture of the tactical situation. The data integrated by Polaris includes that gathered from the main and navigational radars, the inertial navigation system, GPS, AIS and the electro-optical director. It also integrates external tactical data received through a

An overall view of the *Bouchard* class's bridge structure. The combined bridge and combat information centre, with 360° panoramic vision, is equipped with the Naval Group Polaris combat management centre. This presents a tactical picture from the ship's radars and other sensors, many of which are arranged in or on top of the enclosed sensor mast. *(Bruno Huriet)*

A detailed view of an installed enclosed sensor mast on *Piedrabuena*. The conical part of the mast structure conceals the Terma Scanter 6002 surveillance radar within its radar-transparent structure of glass, resin and foam. A small Safran Vigy Engage electro-optical director is mounted on a forward strut immediately above. Moving further upwards, the 'star' on the upper mast structure supports a Wärtsilä navigation radar as well as communications antennae and IFF transponders. A Rhode & Schwarz direction-finding installation is amongst the equipment at the top of the mast. *(Bruno Huriet)*

An enclosed sensor mast is prepared for being lifted into position aboard *Storni*. A Terma Scanter 6002 surveillance radar is housed within the lower, conical part of the mast structure. *(Bruno Huriet)*

Naval Group NiDL tactical data link. In addition to ensuring situational awareness, it is also responsible for performing fire control functions.

The main Scanter 6002 surveillance radar is supplied by Danish company Terma.[7] Optimised to detect small surface targets, it also monitors the low airspace in the ship's vicinity (aircraft flying at altitudes of up to 6,000ft at ranges out to 15 nautical miles) and can be used for helicopter landing control. The system is located in the conical, lower part of the enclosed sensor mast, which is fabricated from radar-transparent material: glass, resin and foam. Its design and manufacture presented a complex challenge for Naval Group, as the structure had to allow radar transmissions whilst being sufficiently robust to support the higher, aluminium part of the mast that is used to carry other sensors and antennae. However, the benefits of having a single weatherproof structure outweighed these difficulties. Other sensors include two Wartsilä navigation radars, identification friend or foe (IFF) equipment and a Safran Vigy Engage electro-optical detection system.[8]

Combat and navigation systems also rely on input from the inertial navigation system – an iXblue Marins N3 – that is based on fibre optic gyroscope technology.[9] It provides a highly accurate position, as well as information on heading, roll and pitch. A Netans data distribution unit interfaces with the ship's systems and sensors.

The class's bridge layout is designed to allow a compact team to perform navigational and operational duties, including surveillance, communications, weapons direction and helicopter control. The forward-most row of bridge consoles integrate the two S (NATO E/F) band and X (NATO I/J) band navigation radars, an electronic chart display and information system (ECDIS), GPS, AIS and a fire alarm and fire extinguishing control panel. The helm station, at the centre, has engine and bow thruster controls at hand. Slightly behind, a second row of consoles incorporate the Polaris combat management system, with screens giving a complete picture of the situation around the ship on the basis of inputs from all the sensors. The main gun, with its electro-optical sensor suite (day and night sensors and laser rangefinder) is controlled from a panel integrated into this position. Moving aft, a global maritime distress and safety system (GMDSS) compliant communications suite is designed for dual military and civilian use and encompasses the usual array of radio links, including satellite communications. The aviation control station is located on the starboard side of the aft-facing windows overlooking the helicopter deck whilst remote control stations for the two machine guns are located on both sides of the bridge.

A close-up of the Leonardo Marlin weapon station equipped with a 30mm gun installed in the main foredeck position on *Piedrabuena*. This system has been specified by the Argentine Navy in lieu of the Nexter 20mm modèle F2 gun that equipped *Bouchard* when she was in *Marine Nationale* service as *L'Adroit*. *(Bruno Huriet)*

Armament: Like most OPV classes, *Bouchard* and her sisters are not intended to become involved in intensive combat operations. However, light weapons are still required to support constabulary missions; in some cases the ARA has had to resort to its weapons to stop foreign fishing vessels that were trying to evade or even ram into its patrol vessels.

In French service, the former *L'Adroit* was armed with a Nexter 20mm modèle F2 gun supplemented by lighter machine guns. However, the ARA has selected a somewhat heavier armament focused on a Leonardo Marlin weapon station equipped with a 30mm gun installed in the main foredeck position. Additional weapons include two Reutech 'Sea Rogue' 12.7mm remotely-controlled machine guns, which are fitted on each of the bridge wings.

A key requirement of modern OPV designs is the ability to accommodate modular systems to support a broad range of operations. The ARA intends to follow this trend through use of a range of standard 20ft equivalent unit (TEU) containers to house equipment ranging from pollution control systems to a decompression chamber for divers. Two containers can be stowed on the helicopter deck with the three purpose-built vessels being equipped with a 2.3-tonne capacity crane with a 10m radius.

Accommodation: The former *L'Adroit* was operated by a core crew of thirty-two whilst on loan to the French Navy, with a further twenty-seven berths available for additional personnel such as inspection teams, trainees or Special Forces. Argentina, which envisages the ships carrying out long-duration patrols in often difficult weather conditions, anticipates a crew of up to forty officers and sailors, leaving spare accommodation for at least nineteen additional mission-specific embarkees.

Accommodation standards are generally high, with ratings accommodated in berthing spaces

L'Adroit pictured towards the end of her time in French Navy service in April 2018. She was subsequently refitted between February and December 2019 prior to joining the Argentine Navy as *Bouchard*. This process involved many equipment changes. *(Bernard Prézelin)*

Storni being towed from the Kership facility at Lorient for outfitting at Piriou's yard at Concarneau in the summer of 2020. The three members of the class built to specific Argentine Navy order are being completed by the Kership joint venture between Naval Group and Piriou. *(Bruno Huriet)*

housing a maximum of ten sailors with adjoining wet areas. The layout allows for a mixed-gender crew with dedicated areas for female personnel. Chief petty officers (CPOs) and officers occupy single or two-berth cabins which are either equipped with en-suite modular bathrooms or nearby wet areas. The commanding officer has a separate wardroom adjacent to his cabin just below the bridge; the executive officer's cabin is also located on the same deck.

A fully-equipped galley serves an officers' wardroom, a CPOs' mess and a ratings' mess room, which are all located in the same vicinity, Dry provision stores and cold rooms provide sufficient food storage to support missions of durations of up to 30 days without replenishment. In order to comply with MARPOL pollution prevention requirements, a dedicated cold store allows garbage to be kept on board before offloading for disposal ashore.

CONSTRUCTION AND DELIVERY

The construction of *Bouchard* and her three sister ships has extended over a period of more than a decade due to the gap between the first project for the self-funded ship and the subsequent Argentinian order for the three additional ships. This has also had the result that their construction phases have been very different.

Construction of the self-funded ship – what was to become the French Navy's *L'Adroit* and is now ARA *Bouchard* – was focused on a dedicated building facility at Lorient that also housed the complete project design and production teams.[10] This allowed for very close collaboration that – combined with the integration of several key suppliers within the project's framework – considerably shortened programme completion. One major component of the vessel that was sub-contracted was the forward half of the hull, which was fabricated by Piriou in Concarneau before being joined to the Lorient-built after structure in the building hall. The completed hull was then transferred into a barge that was then submerged to put the ship afloat. As mentioned previously, the period from first steel cutting to delivery was only in the region of some 18 months.

Piedrabuena is the only ship in the class to have been entirely constructed at the Piriou shipyard in the town, although the facility has been responsible for outfitting all three members of the class that have been built to specific Argentine order. *(Bruno Huriet)*

After six years of operations with the French Navy, a major overhaul was required before the ship's transfer to Argentine Navy ownership. A ten-month refit was completed between February and December 2019 that saw particular attention paid to the propulsion train, with gearboxes, shafts and propellers all replaced. Additional modifications were completed to adapt the ship to the ARA's requirements with respect to areas such as communications, helicopter refuelling and accommodation. In addition some specific French Navy equipment was removed and new systems installed. Interestingly, the 30mm Marlin weapon station was not fitted until after the ship's arrival in Argentina due to the delivery lead time required by the gun's manufacturer in view of the pandemic.

A different industrial structure was implemented for the three follow-on vessels built specifically for Argentina, largely reflecting the intervening formation of the Kership joint venture. This will see the OPVs built at Piriou's shipyard at Concarneau and at Kership's facility at Lanester, near Lorient. More specifically, the first of the new ships (and second of the overall class) – *Piedrabuena* – has been built completely in Concarneau. The plan for the other two ships involves completion of the hulls at Lanester prior to transfer to Concarneau for outfitting.

As already explained with respect to the class's design particulars, significant modifications requested by the Argentine Navy have been specified for the three series-built ships. These include:

- Increased propulsion capacity through the substi-

Another view of *Piedrabuena* departing French shores in May 2021 at the start of her long delivery voyage to Argentina. She subsequently arrived at the Mar del Plata Naval Base south of Buenos Aires on 23 June 2021. Mar del Plata will be the home for all four members of the *Bouchard* class, which will have the primary mission of ensuring the protection Argentina's extensive exclusive economic zone. The fight against illegal, unreported and unregulated fishing will form a major part of their operational assignments. *(Bruno Huriet)*

tution of 16-cylinder ABC engines in lieu of the 12-cylinder DZC series units found in *L'Adroit/Bouchard*, allowing the 20-knot design speed to be sustained at only 80 percent of the engines maximum continuous rating (MCR).

- Provision of a bow thruster to improve low-speed manoeuvrability.
- Installation of shaft-driven generators to increase overall electrical capacity.
- Stabilisation by means of active fin stabilisers rather than the use of passive tanks.
- Changes to the specified armament and communications systems from first build.
- Shipping of a deck crane to facilitate container handling.

As of mid-2021, both *Bouchard* and *Piedrabuena* had been delivered to the ARA, with completion of the programme scheduled for the spring of 2022.

OPERATIONS

During its French Navy career, the former *L'Adroit* carried out a broad range of missions reflective of typical activities carried out by an OPV. These included patrolling the EEZs of France's overseas departments in the Indian Ocean; a two-month assignment with the European Union's 'Atalanta' anti-piracy task force; the evacuation of citizens from Yemen during that country's civil war; participation in several patrols in the Aegean sea to curb illegal immigration; and a campaign of inspections of tuna fishing vessels to ensure compliance with regulatory limitations. As part of its role as a 'sales ambassador' for Naval Group, the ship also made port visits as far apart as Argentina, South Africa, the

Piedrabuena at sea off the coast of Brittany in March 2021. Her sister-ship *Bouchard* – the former *L'Adroit* – has already proven her value in operations aimed at ensuring the integrity of the nation's valuable exclusive economic zone. *(Marc Ottini)*

UAE and Vietnam as well as participating in major naval exhibitions such as IDEX and IMDEX.

In Argentine Navy service, all four OPVs will be based at the Mar del Plata Naval Base, where their primary mission will be to patrol the country's EEZ. Responsibilities will encompass:

- The surveillance and control of natural resources, most obviously fishing stocks.
- The protection and control of maritime traffic.
- Search and rescue missions.
- Humanitarian and disaster relief operations, including the provision of medical assistance in maritime and riverine areas.
- The protection and support of minor naval units.
- Maintaining an Argentinian naval presence.

Amongst these missions, the fight against IUU fishing is a key priority. Fishing fleets of 400 or more vessels – mainly Chinese – supported by their own replenishment vessels have transited the Magellan Strait to plunder waters close to Argentina's 200-mile EEZ limits (the so-called 'Mile 201 Fleet'), often passing into the EEZ to fish. A comprehensive system is being developed by Argentina to monitor these activities, encompassing both airborne patrols and naval units and using AIS and radar surveillance. The challenge is huge given the vast waters to be covered and the fact that some vessels extinguish their AIS and navigation lights!

ARA *Bouchard* was officially incorporated into the *Armada de la República Argentina* at Mar de Plata on 10 February 2020. She had been transferred into Argentinian ownership on 20 December 2019 and subsequently commenced the near 7,000-nautical mile journey from Toulon to her new home in mid-January. In early May 2020, during one of her first missions patrolling Argentina's EEZ, she undertook the initial interception referenced in the introduction, chasing a 62m-long Chinese jigger after the vessel failed to respond to radio messages and flares.[11] Once finally halted, a boarding party deployed using one of *Bouchard*'s RHIBs found around 1,000 tons of fish on the Chinese vessel, which was escorted into Bahia Blanca to face a substantial fine. Doubtless, the ARA anticipates that this will become an increasingly common event as the other OPVs in the class start to be deployed.

Notes

1. Tracing its origins to the navy-controlled dockyards, DCN was established in 2003 when the former *Direction des Constructions Navales* was transformed from a state entity (a *service à compétence nationale* or SCN) into a private limited company under full state ownership. Subsequently, in 2007, the French armaments conglomerate Thales took a 25 percent stake in the group's equity and the company was renamed DCNS. The Thales stake was increased to 35 percent in 2011 whilst the current name, Naval Group, was adopted in mid-2017.

2. The 'Gowind' series of corvettes and light frigates have gained considerable export success, commencing with an order for six enlarged 'Gowind Combat' variants from Malaysia in December 2011. Displacing around 3,000 tonnes and heavily armed, construction of all the ships was allocated to the local Boustead Naval Shipyard, which has subsequently struggled to complete the contract. In 2014, Egypt became the second country to place an order, signing a deal for a total of four 'Gowind 2500' vessels. The first of these, *El Fateh*, was completed at Naval Group's Lorient shipyard and delivered in September 2017. Licensed construction of the other three has been assigned to the Alexandria Shipyard, which handed over *Port Said* – lead ship of its own batch – in January 2021. The United Arab Emirates and Romania also both announced the selection of 'Gowind' corvettes to meet their own requirements in the course of 2019, although finalisation of the latter contract has been delayed due to legal objections from competitors.

3. The Fassmer OPV80 design was first selected by Chile for its *Piloto Pardo* class OPVs, four of which have been completed in two different configurations by the ASMAR yard at Talcahuano between 2008 and 2017. The Colombian Navy also operates three of the type, which were licence-built by its COTECMAR shipyard in Cartagena.

4. More specifically, the design is compliance with Bureau Veritas Rules for Steel Ships as well as the IMO's Code on Intact Stability A749 and its Code of Safety for Special Purpose Ships MSC266. Other usual rules for commercial ships have also been implemented, including the COLREG collision regulations and MARPOL regulations relating to the prevention of pollution.

5. Whilst *L'Adroit* was in French Navy service, a Schiebel S100 Camcopter unmanned aerial vehicle (UAV) was trialled on the ship under the auspices of France's armament procurement agency, the *Direction Générale de l'Armement* (DGA). Deploying a set of electro-optical and infrared sensors, it provides a significant extension to the ship's surveillance abilities. It is understood, however, that the ARA has no immediate plans to deploy UAVs from the class.

6. *L'Adroit* was built with two ABC 12 DZC diesel engines, each producing a lower power output in the region of 2,700 kW.

7. When completed, *L'Adroit* was fitted with an unusual dual surveillance radar combination under which a longer-range Terma 4100 radar shared the same antenna as the shorter-ranged Terma 6002. The ARA decided not to continue with this combination, partly because the lack of an immediate requirement to operate UAVs made the Terma 4100 system's capabilities less essential.

8. *L'Adroit* was fitted with a range of electronic countermeasures/electronic support measures (ECM/ESM) equipment whilst in French Navy service. These, along with physical decoy launchers and equipment needed to communicate with the embarked UAV, were amongst systems removed before the transfer to Argentina. Similarly, the Safran Vigy Engage fire-control system is a substitution for the Sagem EOMS NG optical director initially installed.

9. A SIGMA 40D inertial navigation system was used in *L'Adroit* during her French Navy service. It is believed that this system has been retained in the transferred ship, resulting in another small difference between her and the purpose-built vessels.

10. As *L'Adroit* was not built for the navy it was constructed under the supervision of the French maritime authorities, being registered and carrying an IMO number before being transferred to naval control.

11. Jiggers are fishing vessels that deploy a large number of lines attached to jigger winches from the sides of the boat. In the South Atlantic, they are usually large vessels designed to catch squid, being equipped with high-powered lights to attract the cephalopods to the surface.

12. The author would like to acknowledge the indispensable assistance of the following people involved in the *Bouchard* class project:

Mrs Stéphanie Giroud, Argentine OPV Operations Manager, Naval Group
Capitán de Navío Víctor Rodolfo Ortiz, ARA, Chief of the OPV Construction Supervision Team
Mrs Nadine Rolland, Communications Manager, Piriou

French speakers are referred to the *Mer et Marine* website – www.meretmarine.com/fr – as a source of additional information on both the Argentine OPV programme and the wider 'Gowind' series of vessels.

532

3.2 SIGNIFICANT SHIPS

STEREGUSHCHIY CLASS CORVETTES

Russia's Green Water Combatants for the 21st Century

Author:
Nikolai Novichkov

The Project 20380 *Steregushchiy* class corvettes – together with the Project 20385 *Gremyashchiy* class iteration of the design – are second-tier naval combatants; or small patrol ships according to their Russian classification. They have been designed for both Russian Navy service and also for potential sale on the international market by the St Petersburg-based Almaz Central Naval Design Bureau, which is now a subsidiary of Russia's United Shipbuilding Corporation. Ships of these classes will form the backbone of the Russian Navy's green water component for the next three to four decades.

PROJECT BACKGROUND

The development of the Project 20380 corvette design took place in an environment where the need for such ships was not only reflected in Russia's particular needs but also in broader global trends. The requirement to ensure the security of a state's 'green waters' – including securing naval bases, commercial ports and other coastal infrastructure, as well as protecting economic and security interests across the broader expanses of exclusive economic zones (EEZs) – has become of growing importance in recent decades.[1] The use of small naval platforms, including corvettes, to accomplish these tasks has become increasingly common.

The attraction of corvette-sized combatants has also been influenced by technical considerations. Recent achievements in the field of miniaturising shipborne weapons systems, along with simultaneous improvements in enhancing their effectiveness, have allowed relatively small naval platforms to be equipped with a disproportionate combat capability. This has obvious relevance where political and economic factors have acted as a constraint on naval procurement activity. It is also worth noting that these changes have had an impact on the duties that can be performed by corvette-sized ships, which were historically focused on single missions such as surface-strike or anti-submarine warfare given a limited armament capacity restricted their multi-role capabilities. Today, the integration of powerful artillery systems, new missiles and shipborne helicopters on corvette-sized warships make them true multipurpose platforms. Moreover, a further increase in their future potency is heralded by the integration of robotised sub-systems, such as unmanned aerial, surface and underwater vehicles.

The security of the country's near seas also has a particular importance in Russian defence strategy given the traditional role of naval 'bastions' as safe havens for strategic forces, notably the navy's ballistic missile submarines. The effectiveness of these bastions depends on the possession of strong anti-access/area denial (A2/AD) capabilities, a role very much suited for multi-role corvettes with strong anti-surface and anti-submarine capabilities and operating within the envelope of land-based air superiority. Whilst corvettes are probably too small for fully effective operations in seas patrolled by the Russian Navy's Northern Fleet, it would seem likely that ensuring the security of the Pacific Fleet's bastion in the Sea of Okhotsk was one of the factors influencing the Project 20380 requirement.

Project 20380 Origins: Initially designated 'Korvet-1' during its design stage, the origins of what was to become the Project 20380 corvette can be traced to difficulties with implementing the previous programme for the Project 12441 multi-purpose patrol ship. Just as for the Project 20380 class, the Project 12441 had been devised by the Almaz Central Naval Design Bureau.[2] A prototype vessel – named *Novik* – was laid down at the Yantar shipyard in Kaliningrad in 1997 but was ultimately seen as too sophisticated for the green water corvette role and hence never completed. The Russian Navy (VMF) therefore turned its attention to a less sophisticated – and less expensive – vessel.

The Project 20380 design was intended to perform multi-role operations in the green waters of Russia's near seas, including anti-surface and anti-submarine warfare, the provision of gunfire support for amphibious operations and general patrol duties. This multi-functionality contrasted with the previous single-role optimisation of Soviet-era

Left: The Project 20380 corvette *Boikiy* pictured at St Petersburg in August 2016. Incorporating advanced stealth features and multi-role combat capabilities, the Project 20380 design is intended to ensure the security of Russia's near seas. *(Conrad Waters)*

coastal combatants. Other key design requirements included careful attention to signature suppression and increased levels of automation compared with legacy ships.

The design effort on the new ship was led by Almaz under the leadership of chief designer I N Ivanov. Research support was provided by the First Central Research and Scientific Institute of the Russian Ministry of Defence, with numerous other Russian enterprises also involved in the programme. The completion of the design effort in 2001 allowed transition to the production phase. It seems from contemporary reports that construction of twenty-four vessels was initially envisaged, which were to be split equally between the VMF's four main fleets. The St Petersburg-based Severneya Verf (Northern Shipyard) was to build the corvettes destined for the Baltic Fleet, Black Sea Fleet and Northern Fleet, whilst the Amur Shipbuilding Plant in Komsomolsk-on-Amur was allocated the six ships destined for Pacific Fleet.[3]

Steregushchiy, the first ship of the Project 20380 class, was laid down at Severnaya Verf on 21 December 2001. Launched on 16 May 2006, she was formally commissioned into the VMF on 28 February 2008 after an extensive series of trials.

DESIGN DESCRIPTION

The Project 20380 corvettes have a standard displacement of c. 1,800 tonnes and a full load displacement of c. 2,250 tonnes. Waterline length is 90m – extending to 104.5m overall – whilst waterline beam is 11.1m and maximum breath 13.0m.

Table 3.2.1.

PROJECT 20380 – STEREGUSHCHY PRINCIPAL PARTICULARS

Building Information:	
Laid Down:	21 December 2001 **Launched:** 16 May 2006 **Commissioned:** 28 February 2008
Builders:	Severnaya Verf (Northern Shipyard), St Petersburg
Dimensions:	
Displacement:	1,800 tons standard. 2,250 tons full load displacement.
Overall Hull Dimensions:	104.5m x 13.0m x 3.7m (7.9m at sonar dome). Length at waterline is 90.0m and beam 11.1m.
Equipment:	
Missiles:	2 x quad Uran (SS-N-25) SSMs. 1 x Kortik M (CADS-N-1) AAGM mounting. Igla (SA-18) MANPADS.
Main Guns:	1 x A-190 Universal 100mm. 2 x AK-630 30mm CIWS, light weapons.
Torpedoes:	2 x quad SM-588 324mm torpedo tubes (Paket NK system).
Countermeasures:	4 x ten barrel PK-10 'Smeliy' decoy launchers. TK-25 ECM complex.
Aircraft:	Flight deck and hangar for one Ka-27 series helicopter.
Principal Sensors:	1 x Furke 2 3D surveillance radar. 1 x Monument A target designation radar. 1 x Puma 5P-10 artillery fire-control radar. Navigation radar. 1 x Zarya 2 sonar suite. Minotavr towed array. Anapa M saboteur detection system.
Combat System:	Sigma 20380.
Propulsion Systems:	
Machinery:	CODAD. 4 x Kolomna D49 series diesels in 2 x DDA 12000 propulsion trains. Total output c. 18MW.[1]
Speed:	Maximum speed is c. 26 knots. Endurance at 14 knots is c. 4,000 nautical miles.
Other Details:	
Complement:	Normal crew is 100, including 14 officers.
Class:	*Steregushchiy* (530) was the only member of the class to be completed to this configuration.

Notes:

1. Sources differ as to whether the Kolomna 10D49 (each 3.9MW) or Kolomna 16D49 (each 4.5MW) diesels are fitted to earlier members of the class.

Hull draught of 3.7m extends to as much as 7.9m at the sonar dome. Maximum speed amounts to c. 26 knots, whilst the ship has an endurance of 4,000 nautical miles at an economical speed of 14 knots. The corvette's crew comprises 100 personnel, of whom 14 are officers. Further technical detail is provided in Table 3.2.1.

Hull and Superstructure: The Project 20380 corvettes benefited from a number of innovations in both design and materials that emerged during the late 1990s. These are most evidenced by significant changes in the shape of the hull and superstructure compared with previous vessels. This evolution can be attributed to two pivotal factors, viz. the use of stealth features to reduce the ship's signature and the enhancement of overall design to improve seaworthiness.

Turning first to stealth, the emergence of a holistic approach to the reduction of warships' radar cross section in recent years has seen a shift from the elliptical design patterns that dominated the surface ships of the 1970s to new hull and superstructure shapes that minimise radar reflection. In some conditions, for example, these measures allow the air defensive systems of ships to engage aerial targets before they themselves are detected. The Project 20380 corvettes were designed in accordance with this approach, incorporating:

- Low-silhouette, laconic rectangular superstructures avoiding radar-reflecting angles.
- A carefully designed, flared hull, shaped to

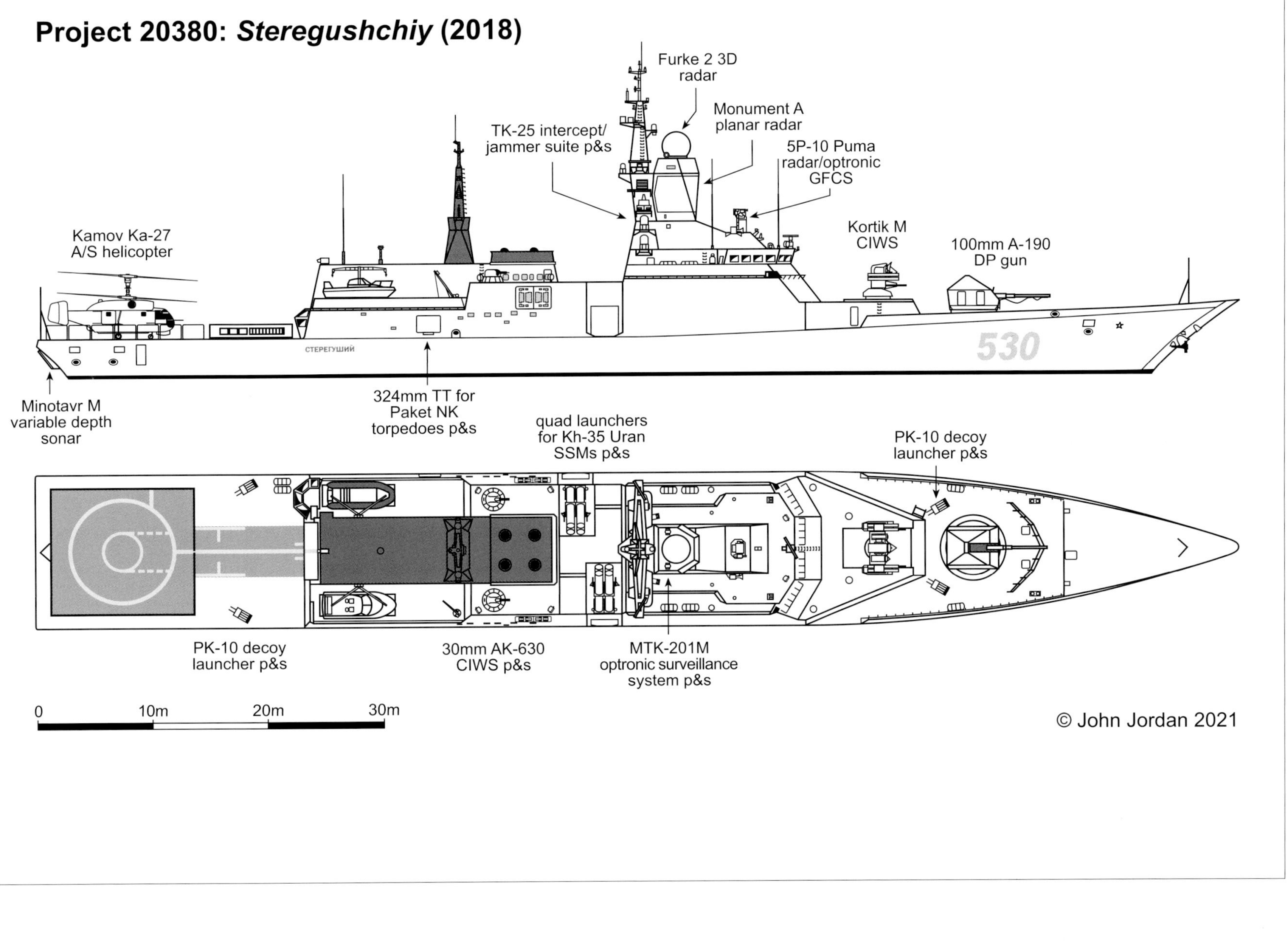

The prototype Project 203800 corvette *Steregushchiy* pictured in June 2009, not long after she was first commissioned. Her stealthy design combines a steel hull with a composite structure of radar-absorbent materials. The 'Kortik M' combined naval anti-aircraft gun and missile system installed forward of the bridge distinguishes her from other early members of the class. *(Nikolai Novichkov)*

An aerial view of the first series-produced Project 20380 class corvette *Soobrazitelniy* at sea in April 2017. The Project 20380 corvettes were one of the first Russian warship designs to benefit from a comprehensive approach to radar stealth, as evidenced by their carefully designed hull and superstructure, as well as the reduction in deck-mounted equipment. *(Crown Copyright 2017)*

connect with the superstructure at an optimum angle.

- Radar-absorbent composite materials and coatings.
- Antennae mounted in a 'stealth mast' transparent only to the ship's onboard systems.
- Under-deck mounted vertical launch systems.
- Retractable weapons systems and antennae.
- Reduction of deck-mounted systems to essential components, all utilising low-reflection shapes.

The introduction of new composite materials in the Project 20380 series has been a particular noteworthy design feature, with much of the ships' superstructure comprised of multilayer composites. According to Nikolai Fedonyuk, the head of the Krylov State Scientific Centre responsible for material strength and composite construction design, several types of radar-absorbent materials with high performance in the relevant electromagnetic spectra were developed. Prominent amongst these new materials are multi-functional, three-layer constructions comprised from carbon-reinforced polymer and fibreglass materials formed with the aid of vacuum technologies.[4]

The use of composite materials has delivered several other benefits. Notably, their use had significant advantages with respect to top-weight, reducing the weight of some assemblies by between 10 and 15 percent compared with previous structures made out of aluminium. According to Fedonyuk, they also increased usable internal volumes by around 5 to 7 percent. Other advantages include the economic benefits of materials that do not require additional painting and are not subject to corrosion. Importantly, as compared with aluminium, they also improve the overall fire resistance and survivability of the ships.

The Project 20380's composite superstructure is combined with a flush-deck hull of traditional steel construction. There are two deck levels below the weather deck level and nine watertight subdivisions. The use of a new underwater hull form is claimed to reduce underwater drag by as much as 25 percent, allowing up to 2 knots of additional speed to be achieved whilst utilising a more compact main propulsion system. The hull form is also more weatherly than in previous ships, allowing the corvettes to remain fully combat effective in conditions up to Sea State 5. As with other recent modern warship designs, a focus on modular architecture

The Project 20380 design is typified by significant use of composite materials to reduce radar signature and improve fire resistance and survivability over previous aluminium structures. This image shows a three-layer composite panel with a corrugated middle layer (front) and a composite panel with U-shaped beams (rear). *(Nikolai Novichkov)*

facilitates the ready integration of new weapons and other combat sub-systems during a service life that is expected to last for at least 30 years and which has already been reflected in ongoing iterative improvements to the class as production has proceeded.

Propulsion: The Project 20380 corvettes are equipped with a twin-shaft combined diesel and diesel (CODAD) propulsion system providing a total power output in the region of 24,000hp (18MW). Each shaft line is equipped with a DDA12000 propulsion train. This incorporates two Kolomna D49 series diesels – each with an output of c. 6,000hp (4.5MW) – that drive a five-bladed, fixed propeller through reverse-reduction gearing. The diesels are reported to have a service life of 50,000 hours before renewal. The design's focus on stealth extends to acoustic considerations, with the design reportedly integrating noise dampening technologies originally developed for Russia's new nuclear-powered submarines. Four DGA-430 diesel generators, each with a power output of 630kW, supply the ship's electrical distribution network.[5]

Combat Suite: The Project 20380's combat suite integrates surface strike, anti-air and anti-submarine weapons systems with radar and other sensors by means of a 'Sigma 20380' combat management system specifically configured for the class. The design's combined bridge and combat management centre are co-located, doubtless helping to minimise total crewing requirements. Many of the principal sensors are also located in the area of the forward

Soobrazitelniy pictured being prepared for floating out at St Petersburg's Severnaya Verf on 31 March 2010. The class's hull form was significantly revised compared with previous designs, with a reduction of underwater drag by as much as 25 percent being claimed. *(Severnaya Verf)*

Anti-submarine detection forms an important part of the Project 20380 design's capabilities. A hull-mounted sonar that forms part of the class's 'Zarya 2' suite is mounted in the bulbous bow, clearly evident in this further image of *Soobrazitelniy* in the course of being launched. *(Severnaya Verf)*

The Project 203800 class squeezes a significant amount of surface strike, anti-air and anti-submarine weapons systems into a compact hull. This August 2016 view of *Boikiy* shows the large A-190 100mm gun mounting in 'A' position, the secondary AK-630 30mm CIWS systems to either side of the funnel and the facilities for helicopter operations aft. The ship's 'Uran' surface-to-surface missile launchers are concealed behind bulwarks between the forward superstructure and funnel whilst quadruple anti-submarine torpedo tubes are hidden behind doors in the superstructure in the area of the hangar. *(Conrad Waters)*

superstructure. The main 'Furke 2' 3D surveillance radar is housed in a prominent radome atop the forward stealth mast. Operating in the E band (US S band), 'Furke 2' is a rotating phased array capable of detecting and tracking targets out to around 150km. The stealth mast also conceals a 'Monument-A' targeting radar for the outfit of surface-to-surface missiles within its composite structure. Additional surface-search and navigation radars are mounted atop the bridge and on side extensions to the after part of the forward mast complex.

An important component of the Project 20380's overall combat capacity is its underwater detection capabilities. A 'Zarya 2' sonar suite includes an antenna that is located in the forward bow bulb. This is supplemented by a 'Minotavr M' towed array deployed from the stern. The design is also equipped with the 'Anapa M' underwater anti-saboteur detection sonar system for protection against divers and similar threats whilst a vessel is stationary.

Electronic warfare capabilities include PK-10 'Smeliy' launchers capable of firing a range of radar and electro-optical decoys. Four ten-cell launchers are fitted, two immediately behind the main gun and two to the sides of the aft helicopter deck. A visually prominent TK-25 jamming complex is also fitted to the sides of the after part of the forward mast structure. Manufacturer's data suggests that the system is capable of detecting up to 256 targets from a catalogue of over 1,000 radars and of suppressing four radars simultaneously.

Communications are taken care of by an automated 'Ruberoid' communications hub that integrates systems covering the usual range of radio frequencies.

As initially installed on the prototype Project 20380 corvette *Steregushchiy*, principal weapons systems comprise:

Anti-Surface Missiles: The ships are armed with the 'Uran' (NATO SS-N-25 'Switchblade') missile system. This incorporates two, four-cell inclined 3S-24 launchers for subsonic Kh-35U cruise missiles with a range of up to 260km. The launchers are mounted amidships behind bulwarks between the forward superstructure and funnel.

Air Defence Missiles: The prototype Project 20380 corvette, *Streregushchiy*, is armed with a single 3M87-1 'Kortik M' (CADS-N-1 'Kashtan') naval

A series of detailed views of the corvette *Boikiy* taken in 2016 showing some of her principal equipment. The ship's forecastle is dominated by an A-190 'Universal' 100mm gun that is capable of firing eighty rounds per minute at ranges of up to 21km. It is controlled by the SP-10-02'Puma' director mounted atop the ship's bridge. The 'Redut' VLS system between the gun and bridge was introduced with effect from *Soobrazitelniy*, the second member of the class, replacing the 3M87-1 'Kortik M' naval anti-aircraft gun-missile system found in *Steregushchiy*. The ship's amidships section is occupied by two quad launchers for subsonic 'Uran' anti-ship cruise missiles, concealed behind bulwarks for stealth considerations. Two AK-630 30mm CIWS to either side the funnel cater for last-ditch defence. Also evident is the mast top dome that conceals the design's 'Furke 2' surveillance radar and the various components of the TK-25 ECM system at the sides of the mast. A targeting radar for the 'Uran' missiles is hidden with the mast's forward, composite superstructure. The aft superstructure is dominated by the hangar for a Ka-27-sized helicopter. A small flying-control position is built into the ship's port side. Also visible are the small alcoves for davit-launched small boats and RHIBs, as well as the hangar roof-mounted OSPV-20380 antennae for helicopter command and control. *(Conrad Waters)*

Starting from the second Project 20380 corvette – and first series-produced unit – *Soobrazitelniy*, the 'Kortik-M' system that equipped *Steregushchiy* has been replaced with the 'Redut' air-defence system. This comprises a total of twelve vertical launch cells that can be equipped with 9M100 short-range and the 9M96 series of medium-range and long-range missiles, extending the air-defence envelope. As can be seem from these 2017 images of *Boikiy,* the change has had only a modest impact on the ship's overall appearance. *(Crown Copyright 2017)*

anti-aircraft gun missile (AAGM) system. This combines eight ready-use 9M311-1 (SA-N-11 'Grison') surface-to-air missiles with a range of up to 10km with two six-barrel AO-18K guns. A storage and reloading system provides storage for a total of thirty-two missiles in launch containers in a below-decks magazine. The system is based on the concept of the missiles being used to engage targets at medium ranges, with the cannon dealing with 'leakers'. Man-portable 9k38 'Igla' (SA-18 'Grouse') can also be shipped for supplemental air defence.

Artillery: The Project 20380 class's main gun is an A-190 'Universal' 100mm naval mounting located in 'A' position. First developed during the 1990s, the gun has a rate of fire of c. 80 rounds per minute. It can engage surface targets out to ranges of c. 21km and aerial targets at c. 15km. The ships also carry two six-barrel side-mounted AK-630 30mm close-in weapon systems (CIWS); one on either side of the funnel. Fire control is by means of a 5P-10-02 'Puma' phased-array director mounted on top of the bridge superstructure. The ships can also be equipped with 14.5mm heavy machine guns and

DP-64 grenade launchers for close-range defence against asymmetric attacks.

Anti-Submarine Systems: Shipboard anti-submarine system capabilities are focused on the 'Paket NK' anti-submarine warfare system which incorporates two trainable quad SM-588 launchers for 324mm torpedoes. One mounting is located to port and starboard abreast the funnel. The system is equipped with anti-submarine and counter-torpedo torpedoes for close range defence against the underwater threat.

Offensive anti-submarine capabilities are largely in the hands of a Kamov Ka-27PL 'Helix-A' anti-submarine warfare helicopter, for which both a flight deck and hangar are provided. A dedicated 'flyco' position is located to the port side of the hangar whilst OSPV-20380 antennae mounted on the hangar roof are used to provide navigation direction whilst the helicopter is in flight. *Streregushchiy* was the first Russian Navy ship of this type to be provided with such extensive helicopter facilities.

DESIGN ITERATIONS

The Project 20380 series of corvettes have been subject to progressive development since the prototype vessel, *Steregushchiy*, was constructed.

Starting from the second, series-production unit, *Soobrazitelniy*, the 'Kortik-M' system has been replaced with the 'Redut' air-defence system. This comprises three four-cell 3S 97 vertical launch modules that can be equipped with 9M100 short-range and the 9M96 series of medium-range and long-range missiles also deployed in Russian land-based surface-to-air missile systems. This improvement provides an extended air-defence 'umbrella' compared with 'Kortik M'.[6] Later members of the Project 20380 series are also benefitting from technological advances incorporated in the Project 20385 *Gremyashchiy* class variant. Most notably, these include a revised forward mast incorporating a new multi-function radar designed by the JSC Zaslon enterprise. The Amur-built *Hero of the Russian Federation Aldar Tsydenzhapov* – the seventh Project 20380 corvette to be completed – was the first of the class to enter service with the new system when she was delivered on Christmas Day 2020. Plans are also reportedly being considered to substitute the universal UKSK vertical launch system (VLS) capable of carrying 'Oniks' (SS-N-26 'Strobile') and 'Kalibr' (SS-N-27 'Sizzler') anti-ship and land-attack missiles for the 'Uran' missile launchers fitted to existing ships in units now under order. The UKSK launcher can also support the Tsirkon (SS-N-23) hypersonic cruise missile that is currently undergoing trials.

The improved Project 20385 *Gremyashchiy* class was developed by Almaz some years after work on the initial Project 20380 design was completed. These are larger ships, with a waterline length of 92m, an overall length of 106.3m and a full load displacement approaching 2,500 tons.[7] The increase in size has allowed the incorporation of an eight-cell 3S-14 UKSK VLS for surface-to-surface missiles. They replace the 3S-24 launchers for 'Uran' missiles but are actually located forward of the bridge in the position formerly occupied by the 'Redut' vertical launchers

Table 3.3.2.

PROJECT 20380 BOIKIY PRINCIPAL PARTICULARS

Building Information:

Laid Down:	27 July 2005 **Launched:** 15 April 2011 **Commissioned:** 16 May 2013
Builders:	Severnaya Verf (Northern Shipyard), St Petersburg

Equipment Details as for *Steregushchiy* except:

Missiles:	2 x quad Uran (SS-N-25) SSMs. 12 x Redut VLS cells for 9M100 and 9M96 series SAMs.

Other Details

Class:	*Soobrazitelniy* (531), *Boikiy* (532) *Stoikiy* (545), *Sovershenniy* (333) and *Gromkiy* (335) have been completed to this configuration.

Project 20380: *Boikiy* (2016)

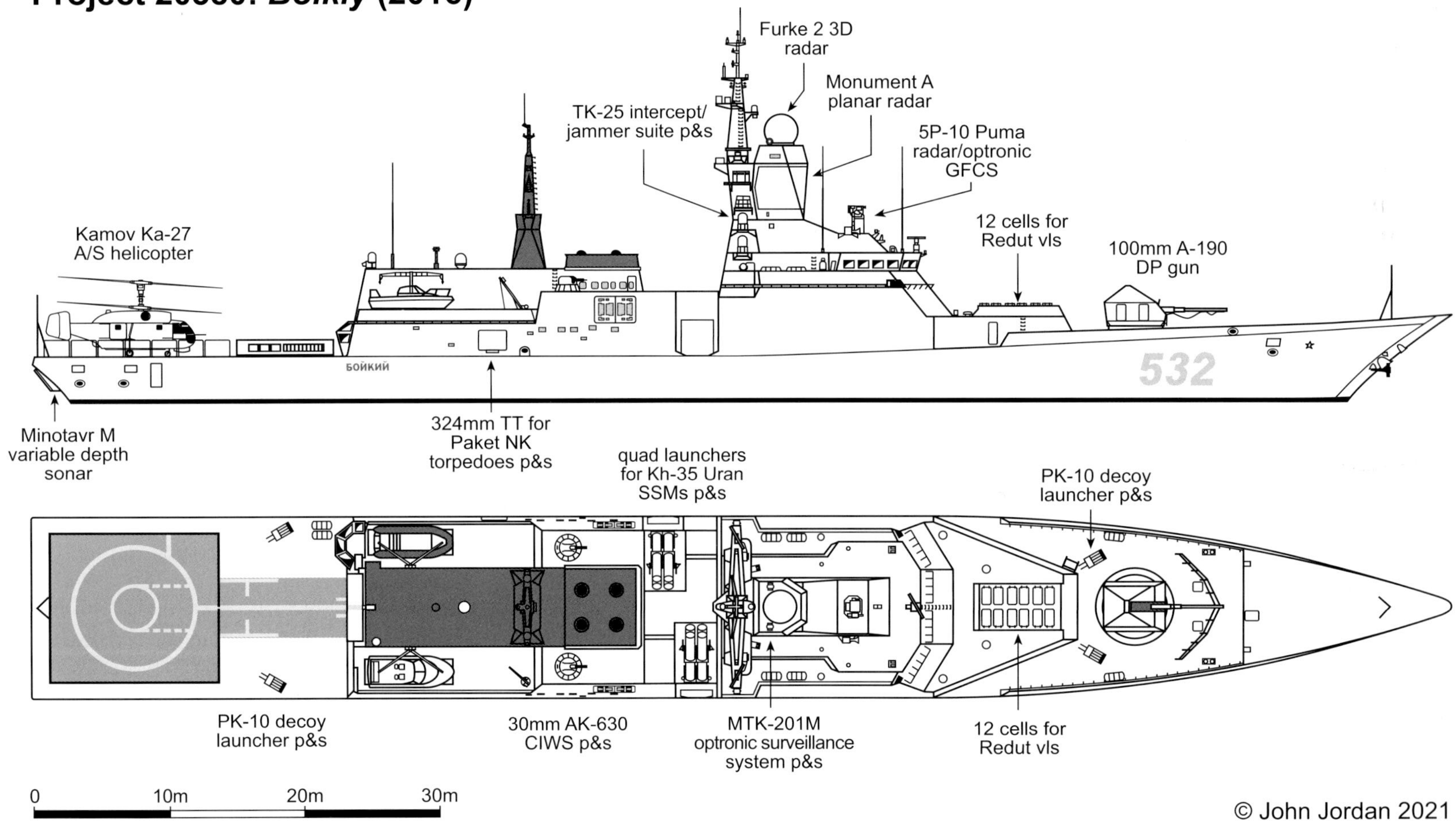

in the Project 20380 class. These are shifted aft to either side of the helicopter deck, gaining an additional four-cell 3S 97 launch module in the process (for a total of sixteen cells). It was also intended that the class would be equipped with a new propulsion system based on German MTU diesel sets but the imposition of sanctions against Russia after the annexation of the Crimea prevented these from being delivered. By this time the two lead ships of the class had already been laid down and considerable delay was encountered until the relevant compartments could be redesigned to accommodate the Kolomna D49 series diesels used in the Project 20380s.

An important aspect of the Project 20385 revisions has been a major enhancement in the electronics outfit most visibly evidenced by a rearrange-

The Project 20385 corvette is an enlarged variant of the original Project 20380 design, incorporating enhanced strike capabilities and a new integrated mast equipped with the 'Zaslon' phased array. These pictures show the lead ship *Gremyashchiy* in the summer of 2019, a few months before she was formally commissioned. *(Nikolai Novichkov)*

Boikiy (left) and *Soobrazitelniy* (right) are seen transiting the English Channel in 2017. The ships form part of an initial group of series-built Project 20380 ships constructed at St Petersburg and in the Far East. Construction has subsequently transitioned to a modified Project 2038 variant, as well as to improved Project 20385 and 20386 types. *(Crown Copyright 2017)*

ment of mast structures and fittings. Notably, the aft mast found in the Project 20380 class is dispensed with and the prominent 'Puma' fire-control system above the bridge deleted. The forward 'stealth mast' has been redesigned to incorporate the 'Zaslon' MF RLK multi-function phased array first fitted in the Project 20380 *Aldar Tsydenzhapov*. Replacing a number of the systems found in the previous class, 'Zaslon' is described as a modular, multi-purpose system comprising a control module; an I/J Band (US X band) 3D radar with fixed active phased arrays; an E Band (S Band) rotating active phased array; electric support measures (ESM) and electronic countermeasures (ECM) modules; and an artillery fire-control system. The system is designed to search for, detect and track aerial and surface targets of all types and sizes as well as performing fire-control functions for both missiles and artillery. It is claimed that it can track up to fifty targets at ranges out to 300km in active mode and at altitudes of up to 30km. Moreover, it is able to detect an aerial target with a reflective area of at least 1m^2 at no less than 75km. A larger number of targets can be tracked in passive mode at ranges of up to 200km.

A computer-generated image of the new Project 20386 corvette, which has evolved from the Project 20380/20385 series. One vessel – named *Merkuriy* – is currently under construction at Severnaya Verf. *(Almaz Central Design Bureau)*

Overall, the panoramic view provided by systems such as 'Zaslon' provides a significant advantage over legacy equipment, enabling the tasks of several radio-technical systems to be performed simultaneously and allowing the ship's crew to assess the situation around them in real time. Its capability is reinforced by an enhanced variant of the 'Sigma' combat management system with fully automated work stations and a revised command centre layout.

A further development of the design is represented by the Project 20386 iteration. The first of these ships was laid down at Severnaya Verf as *Derzkiy* in October 2016 but was subsequently renamed *Merkuriy* in April 2019. She was subject to technical launch in March 2021 and is currently in the process of fitting out. The ship has the same multi-role capabilities of the previous Project 20380/20385 series, being able to attack enemy surface combatants with long-range cruise missiles, prosecute underwater contacts detected by her onboard anti-submarine warfare equipment, perform air-defence functions and support amphibious operations. However, she represents a further increase in size compared with her predecessors and is more capable of blue water operations. It is therefore possible to see her as more of a completely separate class evolved from her predecessors as opposed to being an integral part of the Project 20380/20385 design series.

Merkuriy is reported to have a full load displacement of c. 3,400 tons. She is 109m long and has a beam of 13m. Published images and models of the ship reveal that she will have a monolithic, central superstructure block, which – like the Project 20380/20385 designs – will be manufactured from composite materials. In contrast to her diesel-powered predecessors, she is to be equipped with an acoustically stealthier combined diesel-electric or gas

The first four Project 20380 corvettes to be completed were assigned to the Baltic Fleet but have undertaken deployments to the North Atlantic, Mediterranean and Indian Ocean. These views of *Boikiy* in April 2017 suggest that the ship is making relatively heavy going in the prevailing sea conditions, emphasising the green water nature of the design. *(Crown Copyright 2017)*

(CODLOG) propulsion plant that integrates two 27,500hp (20.5MW) M90FR gas turbines for higher speed operation with two 2,200hp (1.6MW) electric motors for lower speed loitering.[8] Maximum speed will be in the region of 30 knots. The new corvette is also reported to benefit from a modular design approach that includes open architecture combat management systems to allow for the easier integration of new armament and senor packages, including robotised devices. A modular mission bay under the aft part of the flight deck has been provided to support such systems and could also be potentially used to house containerised 'Kalibr' missiles.

Much of the Project 20386's sensor and armament package is drawn from the previous ships. Notably, the 'Zaslon' multi-functional radar complex first devised for the Project 20385 design is integrated into the superstructure. The combat management system will be an upgraded variant of the Sigma system and other anti-submarine and electronic systems are also carried over from the earlier designs. Weapons include sixteen 'Redut' VLS cells mounted in the bow, an A-190-1 100mm gun slightly further aft and launchers for 'Uran' surface-to-surface missiles and 'Paket NK' anti-submarine torpedoes in the superstructure. Close-range defence is supported by two AK-630M 30mm CIWS. A Ka-27 series helicopter can be accommodated in an onboard hangar accessed from a lift in the flight deck.

In January 2020, a scale model of a Project 20386 corvette variant equipped with the UKSK VLS was showcased at an exhibition held at Sevastopol during a visit by Russian President Vladimir Putin. This indicates that Almaz continues to evolve the design and could potentially suggest the configuration of any future, series-built ships. The improved design was notable for its revised and extended bow arrangement, with the 'Redut' VLS launchers moved forward of the main gun to make space for two, eight cell UKSK modules for 'Kalibr', 'Oniks' and, potentially, 'Tsirkon' missiles. This rearrangement provides the variant with significantly greater capabilities than the baseline Project 20386 corvette, providing much greater strike potential as part of a naval deterrence system.

CONSTRUCTION AND ENTRY INTO SERVICE

Plans for construction of the Project 20380 corvettes have been subject to ongoing change since *Steregushchiy* was laid down in St Petersburg in December 2001. Assembly was initially concentrated on Severnaya Verf, which laid the keels for three follow-on, series-built ships between 2003 and

Table 3.2.3: CLASS LIST: PROJECT 20380 AND PROJECT 20385 CLASS CORVETTES

PROJECT 20380 CLASS

NAME	PENNANT	BUILDER	FACTORY No	LAID DOWN	LAUNCHED	COMPLETED	FLEET	NOTES
Steregushchiy	530	Severnaya Verf	1001	December 2001	May 2006	February 2008	Baltic	Prototype with 'Kortik M'AAGM
Soobrazitelniy	531	Severnaya Verf	1002	May 2003	March 2010	October 2011	Baltic	Equipped with 'Redut' VLS
Boikiy	532	Severnaya Verf	1003	July 2005	April 2011	May 2013	Baltic	
Stoikiy	545	Severnaya Verf	1004	November 2006	May 2012	July 2014	Baltic	
Sovershenniy	333	Amur Shipbuilding Plant	2101	June 2006	May 2015	July 2017	Pacific	
Gromkiy	335	Amur Shipbuilding Plant	2102	April 2012	July 2017	December 2018	Pacific	
Retiviy	N/A	Severnaya Verf	1007	February 2015	March 2020	[2021]	–	Equipped with 'Zaslon' MFR
Strogiy	N/A	Severnaya Verf	1008	February 2015	[2021]	[2022]	–	
Aldar Tsydenzhapov	339	Amur Shipbuilding Plant	2103	July 2015	October 2019	December 2020	Pacific	
Rezkiy	N/A	Amur Shipbuilding Plant	2104	July 2016	[2021]	[2021]	Pacific	

A further two Project 20380 corvettes have been ordered from the Amur Shipbuilding Plant and more are expected to be built at Severnaya Verf

PROJECT 20385 CLASS

NAME	PENNANT	BUILDER	FACTORY No	LAID DOWN	LAUNCHED	COMPLETED	FLEET	NOTES
Gremyashchiy	337	Severnaya Verf	1005	February 2012	June 2017	December 2020	Pacific	'Zaslon' MFR & UKSK VLS
Provorniy	N/A	Severnaya Verf	1006	July 2013	September 2019	[2022]	[Pacific]	

A further four Project 20385 corvettes have been ordered from the Amur Shipbuilding Plant and more are expected to be built at Severnaya Verf.

20380 and 20385 corvettes – along with the larger Project 22350 *Admiral Gorshkov* class frigates – are driving the modernisation of the Russian surface fleet. *(Crown Copyright 2017)*

2006. These vessels – *Soobrazitelniy*, *Boikiy* and *Stoikiy* – were commissioned between October 2011 and July 2014. Along with *Steregushchiy*, all these ships are operated by the Russian Navy's Baltic Fleet. They have, however, conducted several long-distance cruises that have involved some units operating in the Atlantic and Mediterranean, as well as even venturing as far as the Indian Ocean.

This initial series of four ships was followed by construction of two additional Project 20380 corvettes at the Amur Shipbuilding Plant in the Far East. One – *Sovershenniy* – was laid down in June 2006 but not commissioned until July 2017. Her sister – *Gromkiy* – was laid down in April 2012 and entered service in December 2018. Both have been outfitted with the same suite of equipment as the three initial, series-built Project 20380s completed in St Petersburg. They serve in the Russian Navy's Pacific Fleet and have also undertaken lengthy voyages, including a cruise to the Asia-Pacific region in May 2021.

The intended shift to construction of the Project 20385 design then led to a pause in Project 20380 construction until 2015-16, when two additional pairs of ships were laid down by Severnaya Verf and the Amur Shipbuilding Plant. This presumably reflected a desire to maintain construction whilst the Project 20385 design was revised to incorporate indigenous systems in place of those that had been subject to sanctions. Of these, only the Amur-built *Aldar Tsydenzhapov* has been delivered to date, joining the Pacific Fleet. As previously mentioned, she is the first Project 20380 variant to be fitted with the 'Zaslon' MF RLK phased array. News reports suggest her Amur-built sister, *Rezkiy* and the Severnaya Verf built *Retiviy* will be completed by the end of 2021. The other Severnaya Verf unit, *Strogiy*, should follow in 2022.[9] This will take the class up to ten completed units.

Turning to the Project 20385 class, work on the first ship – *Gremyashchiy* – commenced early in 2012 at Severnaya Verf but she was only commissioned at the end of 2020. She is due to serve with the Pacific Fleet and should be joined by her sister – *Provorniy* – by 2022. Work also continues on *Merkuriy*, the sole Project 20386 class vessel ordered to date, at the St Petersburg facility. Some reports suggest she will be delivered before the end of 2022 but these would seem to be optimistic.

Looking to the future, in the summer of 2020 it was reported that the Russian Ministry of Defence had determined that additional Project 20380 and 20385 class corvettes would be acquired. Subsequently, in December 2020, it was announced that the Amur yard had received a contract for a total of six additional corvettes – two Project 20380s and four Project 20385s. Work on the first two ships is scheduled to commence before the end of 2021 and all will be delivered for service in the Pacific Fleet by the end of the decade. There have been various reports that Severnaya Verf will also be contracted to build additional vessels, with some sources suggesting the construction of two more of the Project 20380 design and an additional pair of Project 20385 vessels is likely. There has been no official explanation as to why it has been decided to continue production of both variants given the more recent design has now passed trials.

CONCLUSION

It can be seen that the Project 20380 and Project 20385 classes are playing an increasingly important part in the VMF's green water operations. Although introduction of meaningful numbers of the vessels was initially a little protracted, seven Project 20380 and one Project 20385 class corvettes have now been delivered, being split evenly between the Baltic and Pacific Fleets. Additional ships in the course of construction will more than double this number before the end of the decade, potentially seeing ships of the type also joining Russia's other major fleets. Moreover, the decision to place further orders suggests official confidence that the series of designs has met intended requirements. Along with the larger, blue water Project 22350 *Admiral Gorshkov* class frigates, Russia's new corvettes will form the core of the VMF's surface fleet over the coming decades, allowing the majority of the remaining Soviet-era 'legacy' designs to enter a long overdue retirement.

Notes

1. A state's 'green waters' are generally regarded as its near seas, extending beyond the confined 'brown waters' of the littoral up to the open 'blue waters' of the world's oceans.

2. Established in October 1949 as Special Design Bureau 5 (SKB-5), the Almaz Central Naval Design Bureau was originally focused on the development of long-range torpedo and, subsequently, missile boats. The bureau took its current name in 1966. Amongst other important designs originating from the bureau were the Project 1234 (NATO 'Nanuchka') small missile ships, the Project 1241 ('Tarantul') large missile boats and the Project 12322 'Zubr' ('Pomornik') hovercraft. In line with the reorganisation of the Russian shipbuilding industry commenced by President Vladimir Putin in 2007, Almaz now forms part of the United Shipbuilding Corporation (USC).

3. Just as for Almaz, Severneya Verf and the Amur Shipbuilding Plant are now both part of Russia's USC.

4. It is believed that similar materials have been used in fabricating structures for the Project 12700 'Alexandrite' class mine countermeasures vessels.

5. The Kolomna D49 series of medium-speed diesel engines was originally developed for use in Russian railway locomotives but has been used in other Russian warships. As there is some discrepancy in sources as to which precise model of the V16 cylinder D49 variant equips the Project 20380 corvettes, power output figures are somewhat speculative.

6. The 'Redut' system was not, apparently, integrated with a dedicated fire-control system when incorporated into the Project 20380 design, taking targeting data from the 'Furke 2' surveillance radar. There have been reports that this hindered longer-range anti-air engagements. This will be rectified by the installation of the Zaslon-supplied phased array in later ships.

7. From an analysis of published photographs, the increase in length appears to have been achieved by extending the hull in the region of the helicopter flight deck.

8. The M90FR gas turbine was developed by PJSC UEC-Saturn for later members of the Project 22350 *Admiral Gorshkov* class following the interruption of supplies of DA91 gas turbines. It is understood that the third Project 22350 frigate, *Admiral Golovko*, will be the first ship equipped with the new turbine.

9. According to the accounting reports of Severnaya Verf, the price tag for the Project 20380 corvettes *Retiviy* and *Strogiy* was RUB17 billion (c. US$230 million) for each platform. This seemingly low figure appears unlikely to cover government-furnished equipment.

P233

3.3 SIGNIFICANT SHIPS

Author:
Conrad Waters

'RIVER' CLASS BATCH 2 OPVS

The Royal Navy's new 'Colonial Gunboats'

On 7 January 2021, a short service at Portsmouth Naval Base saw the White Ensign hoisted over the new offshore patrol vessel HMS *Spey* for the first time. The fifth and final member of the Royal Navy's Batch 2 'River' class OPVs, she was formally commissioned into the fleet at a ceremony alongside the Ross-shire town of Invergordon on 18 June 2021 prior to her planned deployment to the Far East later in the year. Three of her sister-ships are already serving in various locations across the world under a new strategy aimed at bolstering the Royal Navy's international presence as part of the post-Brexit 'Global Britain' mantra. The fourth, *Tamar*, will also shortly depart for the Far East, the two OPVs becoming the first Royal Navy warships permanently based in the region since the withdrawal from Hong Kong in July 1997. Some have compared the Batch 2 'River' class's new role as being a refreshed version of the 'presence' missions performed by the sloops and gunboats of the British imperial era.

CLASS ORIGINS

The Batch 2 'Rivers' are the latest iteration of a series of OPV designs first developed by the former Vosper Thornycroft shipbuilding group towards the end of the 1990s. Three of these ships – the first batch of 'River' class OPVs – were acquired for Royal Navy service under an innovative leasing and support contract signed in 2001. These 80m-long vessels were followed by *Clyde*, a slightly modified variant equipped with a flight deck for helicopter operations

Left: The 'River' class Batch 2 OPV *Tamar* seen whilst working up in British waters in January 2021. Along with her sister *Spey*, she is set to deploy to the Far East in the course of 2021 as part of emerging Royal Navy plans to maintain an international forward maritime presence. *(Crown Copyright 2021)*

The Batch 2 'Rivers' are the latest iteration of a series of OPV designs originally developed by the former Vosper Thornycroft (VT) shipbuilding group. They are directly derived from a VT 90m OPV class first ordered by Trinidad and Tobago but subsequently acquired by Brazil. This is *Araguari*, the former *San Fernando*. *(Conrad Waters)*

The Batch 2 'Rivers' were constructed at Govan on the Clyde before being loaded out onto and subsequently floated off a submersible barge prior to final outfitting at Scotstoun. This photo shows the lead ship, *Forth,* entering the water in August 2016. *(BAE Systems)*

and intended for service in the Falkland Islands. She was ordered in 2005 under similar contractual arrangements to those agreed for the first three ships. Subsequently, in 2007 an enlarged 90m Vosper Thornycroft OPV design incorporating much of the design philosophy of the Royal Navy vessels but utilising a different hull form was selected by the Trinidad and Tobago Coast Guard. The three ships ordered under this agreement were ultimately acquired by Brazil as the *Amazonas* class when the initial deal collapsed. This 90m OPV design has also formed the basis for the Royal Thai Navy's *Krabi* class. To date, two of these vessels have been completed under licence by the Bangkok Dock Company.[1]

The decision to acquire a second batch of 'Rivers' for British service traces its origins to the long and complex negotiations for the Royal Navy's Type 26 frigate design, also known as the Global Combat Ship. Although a design contract for these frigates had been placed in 2010, the high cost of the programme in an age of austerity meant that an actual construction award was subject to progressive delay. One consequence was a growing likelihood that a significant gap would emerge in the British shipbuilding sector's workload as fabrication of the *Queen Elizabeth* class aircraft carriers' constituent blocks started to reach its conclusion. As well as having negative implications for shipyard employment and employee skills, commercial agreements with Britain's dominant shipbuilder, BAE Systems, meant that large payments would be required simply to keep the shipyards idle.[2]

In November 2013, a way forward was announced. BAE Systems' surface shipbuilding facilities would be rationalised to focus on the Govan and Scotstoun yards on the River Clyde, where three new OPVs would be built to bridge the gap between *Queen Elizabeth* class and Type 26 frigate construction. It was stated that the new patrol vessels would be 'based on a more capable variant of the River class and including a landing deck able to take a Merlin helicopter'.[3] It quickly emerged that these vessels would essentially be the 90m Vosper Thornycroft (now BAE Systems) OPV design, revised to incorporate specific Royal Navy requirements. The need to commence construction relatively quickly to avoid the looming hiatus in shipbuilding activity inevitably had a major influence on the existing design's selection for the new order.

CONSTRUCTION

The contract for what were initially three Batch 2 'River' class OPVs was formally signed with BAE Systems in August 2014. The value of the award was an eye-catching £348m (c. US$490m), being significantly distorted by costs needed to maintain industrial capacity for the future Type 26 frigates. The use of a proven design and previous funding of long-lead items allowed construction to commence relatively rapidly. Steel for the first vessel – named *Forth* – was cut at a ceremony at Govan on 10 October 2014. Start of work on the two following ships – *Medway* and *Trent* – followed in the course of 2015. Orders for two additional Batch 2 'Rivers' – *Tamar* and *Spey* – were placed in December 2016 as a result of decisions taken in the previous year's Strategic Defence & Security Review.[4] The £287m (c. US$410m) contract encompassed unspecified support arrangements for all five ships and took the total value of the programme to BAE Systems to £635m (c. US$900m)

A major underlying objective of the Batch 2 programme was to develop the team that would eventually take forward Type 26 construction, including learning lessons that would refine the approach taken to completing these much more complex ships. The overall build strategy involved initial fabrication and assembly at the Govan yard. Here the use of load-out by means of a submersible barge – claimed to be the first time this had been undertaken for a complete ship on the River Clyde – allowed the ships to enter final outfitting at a higher level of completion than previous vessels, such as the Type 45 destroyers. This activity was largely allocated to the downstream Scotstoun yard, which was to be the centre for work on completion, systems commissioning and final acceptance.[5]

At first, progress with what was a relatively relaxed building schedule proceeded largely to plan. *Forth* was floated out on schedule in August 2016 prior to her official christening at Scotstoun on 9 March 2017. Sea trials commenced in August of that year and she was officially accepted on 25 January 2018, subsequently departing the Clyde for her new home base at Portsmouth. However, significant construction defects – including serious faults with the elec-

Forth being welcomed into her home base of Portsmouth for the first time on 26 February 2018. Although formally commissioned in April of that year, the emergence of construction defects was to delay her entry into operational service. *(Conrad Waters)*

trical distribution system – emerged shortly after her commissioning on what might, with hindsight, be seen as the rather inauspicious date of Friday, 13 April 2018. Significant remedial work delayed her entry into operational service until the following year and also had a knock-on effect on completion of the remaining ships. The second ship, *Medway*, finally commissioned in September 2019, with *Trent* and *Tamar* following in 2020. Whilst a serious embarrassment at the time, the problems confirmed the wisdom of using OPV construction to iron out potential snags before commencement of the Type 26 frigate programme.

Key construction dates are set out in Table 3.3.1.

DESIGN DESCRIPTION

The Batch 2 'River' class OPVs are essentially derivatives of the earlier Brazilian *Amazonas* class. They therefore utilise the same hull form, superstructure and propulsion system of the original Vosper Thornycroft 90m OPV design. However, they also incorporate large numbers of detailed modifications to comply with Royal Navy standards.

A total of twenty-nine specific design changes from the Brazilian ships have been publicly reported, many in the areas of safety and survivability. Amongst the most prominent of these are modifications to watertight integrity and fire security, as well as the incorporation of enhanced firefighting facilities and magazine protection. The electrical distribu-

Table 3.3.1: RIVER CLASS BATCH 2 – CLASS LIST

NAME	PENNANT	ORDERED	FIRST STEEL CUT	LAUNCHED[1]	NAMED[1]	ACCEPTED	COMMISSIONED
Forth	P222	12 August 2014	10 October 2014	20 August 2016	9 March 2017	25 January 2018	13 April 2018
Medway	P223	12 August 2014	8 June 2015	23 August 2017	20 October 2017	5 March 2019	19 September 2019
Trent	P224	12 August 2014	7 October 2015	20 March 2018[2]	13 March 2018	30 October 2019	3 August 2020
Tamar	P233	8 December 2016	8 December 2016	10 October 2018	21 March 2019	25 February 2020	17 December 2020
Spey	P234	8 December 2016	21 April 2017	19 June 2019	3 October 2019	30 October 2020	18 June 2021

Notes:

1. The launch date relates to the date the ship was floated out from her launching barge whilst the named date refers to the date of the official naming ceremony.
2. Uniquely amongst the class, *Trent* was named before her technical 'launch'.

tion system for domestic services has been changed from 115v to 230v and there are improvements to refrigeration and sewage treatment facilities. A scaled-down iteration of the navy's standard BAE Systems CMS combat management system is hosted on the company's new Shared Infrastructure platform. There are also changes to the armament fit and to communications equipment. Many of the most significant changes relate to helicopter operations. These include installation of an enclosed flying control position and flight deck landing grid, as well as modifications to helicopter refuelling arrangements and provision for inflight helicopter refuelling. The flight deck has been adapted to support a heavy helicopter of Merlin size.

Whilst still being optimised for constabulary operations, the sum of these changes has been to make the Batch 2 'Rivers' far more robust and sophisticated ships than the more basic OPVs operated by many navies. An overview of key design features follows.

Hull and Platform Systems: The Batch 2 'Rivers' are c. 90m in length, have a maximum beam of c. 13.5m and displace in the region of 2,000 tons. Their hull form is derived from the experience that the former Vosper Thornycroft gained constructing a series of corvettes and light frigates for export from the 1960s onwards. This design provenance may well account for their comparatively high maximum speed of around 25 knots, which is somewhat greater than that typically found in vessels of this type.

THE ROYAL NAVY'S NEW OFFSHORE PATROL VESSELS

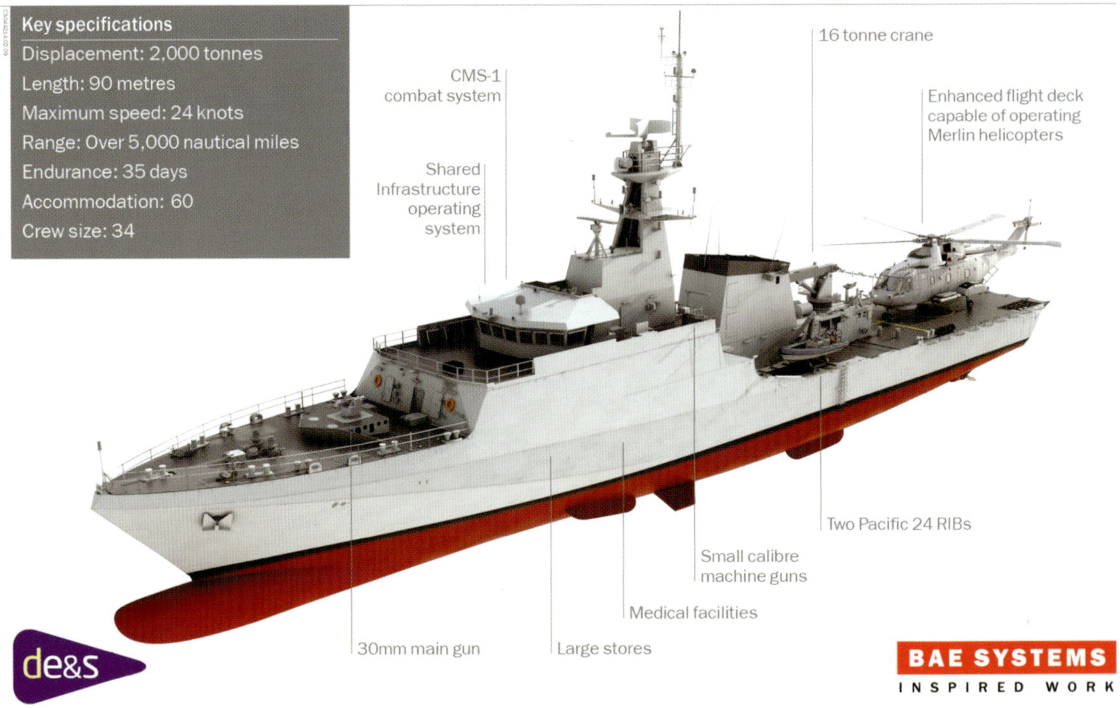

A BAE Systems graphic showing some of the main features of the Batch 2 'River' Class OPV design. The new ships incorporate twenty-nine specific design changes from their Brazilian predecessors, amongst the most prominent of which are significant revisions to helicopter operating arrangements. *(BAE Systems)*

The design's overall configuration is marked by a prominent forecastle deck that breaks to the upper deck just aft of the forward superstructure block. The forward part of the ship is taken up with the magazine and storage spaces, whilst accommodation and messing areas are concentrated in the superstructure and hull decks below the bridge. The midships part of the design is dominated by the machinery spaces, including a large engine room for the twin main diesel engines. Moving aft, there are facilities for an embarked military force immediately beneath the flight deck, with fuel tanks below.

The Batch 2 'Rivers' are powered by two MAN 16V 28/33D medium-speed diesels.[6] Each drives a shaft-line equipped with a Wärtsilä CPP 5C10 controllable pitch propeller through a single input/output, horizontal offset reduction gear. A bow thruster is also fitted to assist low-speed manoeuvrability, whilst a transom flap assists overall propulsive efficiency. In addition to the high top speed referenced above, the class have a respectable c. 5,500 nautical mile range at an economical cruising speed of 12 knots. Stability in adverse weather conditions benefits from the provision of bilge keels and active fin stabilisers. The design has

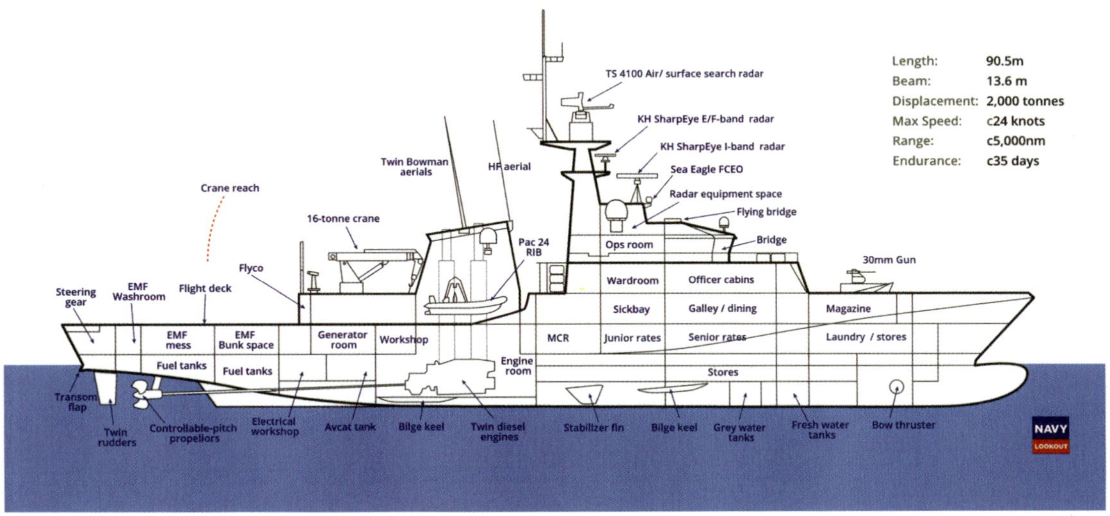

An internal general arrangement drawing of the 'River' class Batch 2 design. *(Courtesy Navy Lookout – www.navylookout.com)*

Above: *Medway* pictured at speed during operations in the Caribbean. The class has a maximum speed in the region of 24–25 knots and an endurance at cruising speed in excess of 5,500 nautical miles. *(Crown Copyright 2020)*

Right: An unusual overhead view of *Forth* berthed at the Mare Harbour facility in the Falkland Islands. The main 30mm gun is mounted on the prominent forecastle forward of the bridge superstructure block, with accommodation and messing facilities concentrated in the superstructure and hull decks below the bridge. The ship's midships section is dominated by the outlets for the twin diesel engines, with the crane and positions for two standard containers immediately aft. There is additional accommodation for an embarked military force immediately beneath the large helicopter deck. *(Crown Copyright 2020)*

proved to be highly agile in the course of trials, doubtless aided by the specification of twin rudders. Reports at the time the *Amazonas* class were constructed indicated that the basic design had a turning circle of three and a half ship lengths and could stop in less than four ship lengths at high speed.

Machinery, electrical distribution, heating ventilation and air-conditioning (HVAC) and other auxiliary services are monitored and controlled by means of a Servowatch integrated platform management system (IPMS) hosted on the Shared Infrastructure platform. As is increasingly common in modern warships, the IPMS allows the machinery spaces to be entirely unmanned during normal cruising operations, with control typically exercised from a console on the bridge. A dedicated machinery control room (MCR) located just forward of the main engine room is manned as required, also serving as the principal HQ1 damage control centre. Extensive bridge automation, including integration of an OSI warship electronic chart display and information system (WECDIS) and warship AIS (WAIS), allows bridge operations to be watched over by as few as three personnel.

Sensors and Armament: Although the Batch 2 'Rivers' are essentially equipped for constabulary operations, the use of an iteration of the standard BAE Systems CMS installed across much of the navy's front-line combatants provides relatively sophisticated command and control capabilities. Their principal sensor integrated into the CMS is a Terma Scanter 4100 surveillance radar. A 2D radar operating in the X band (NATO I/J bands), it performs air and surface-search functions out to a radius of around 110 nautical miles (c. 205km). Capable of supporting helicopter and unmanned aerial vehicle (UAV) control functions, its high resolution makes it particularly well suited to spotting the very small targets that can often form part of an asymmetric threat. The radar suite is completed by a

Table 3.3.2.

HMS FORTH PRINCIPAL PARTICULARS

Building Information:	
Fabrication Commenced:	10 October 2014 **Launched:** 20 August 2016[1] **Delivered:** 25 January 2018
Builders:	BAE Systems at its Govan & Scotstoun yards
Dimensions:	
Displacement:	2,000 tons full load displacement.
Overall Hull Dimensions:	90.0m x 13.5m x 3.5m
Equipment:	
Armament:	1 x MSI Defence Systems DS30M remotely-controlled 30mm cannon. Light machine guns.
Aircraft:	Flight deck for 1 x 15-ton helicopter (no hangar).
Principal Sensors:	Terma Scanter 4100 surveillance radar. Navigation radar. Chess Dynamics Sea Eagle FCEO electro-optical fire-control system.
Combat System:	BAE Systems CMS combat management system. Integrated communications suite.
Other:	2 x 8m RHIBs for boarding operations. Positions for two TEU containers. Up to four additional containers can be stowed on the flight deck.
Propulsion Systems:	
Machinery:	Diesel. 2 x MAN 16V 28/33D diesels each rated at 7.4MW. Two shafts. Bow thruster.
Speed:	Maximum speed is c. 25 knots. Endurance is in excess of 5,500 nautical miles at 12 knots.
Other Details:	
Complement:	Normal crew is c. 50. Accommodation is provided for c. 60–70 crew plus a further 50 personnel in austere accommodation.
Class:	Five Ships: *Forth* (P222), *Medway* (P223), *Trent* (P234), *Tamar* (P233), *Spey* (P234). Three similar ships are in service in Brazil and two in Thailand.

Notes:

1. Refers to technical launch from a floating barge.

pair of Kelvin Hughes SharpEye arrays operating in the NATO I and E/F bands and primarily used to support navigation functions.[7] The military communications system is supplied by Rhode & Schwarz. This supports a wide range of conventional radio and satellite links across the various frequency bands.

The class's principal armament is a single DS30M Automated Small Calibre Gun (ASCG) mounting incorporating a 30mm Bushmaster II cannon located just forward of the main superstructure. The weapon is directed by means of an off-mount Chess Dynamics Sea Eagle FCEO electro-optical director and controlled from a remote console. Its principal role is to provide self-defence against fast inshore craft and similar asymmetric threats out to ranges of around three nautical miles. Positions are also provided for two 7.62mm Miniguns and four general purpose machine guns.

The nature of an OPV's typical duties means that interdiction and search capabilities form a major part of its equipment. The Batch 2 'Rivers' typically embark two davit-launched 'Pacific 24' rigid-hulled inflatable boats (RHIBs) for this purpose, with one each being located alongside the funnel in each of the port and starboard waists. Pictures of the class on deployment have also shown the class is able to c arry other light boats, including larger inflatables and Rigid Raider assault craft. These are seemingly deployed by means of the ship's crane.

Although the Batch 2 'Rivers' can land and refuel all the helicopters currently in the Royal Navy's inventory, the lack of a hangar to support sustained rotorcraft operations has been a criticism frequently levelled at the class. To some extent, this absence is mitigated by the ability to both store and handle containerised equipment which could, for example, facilitate the housing and maintenance of a range of unmanned systems. There are two dedicated positions aft of the RHIB davits for standard 20ft equivalent unit (TEU) containers. Additional containers could be stowed on the flight deck if required. Each ship is fitted with a Pellegrini crane that has a 16-

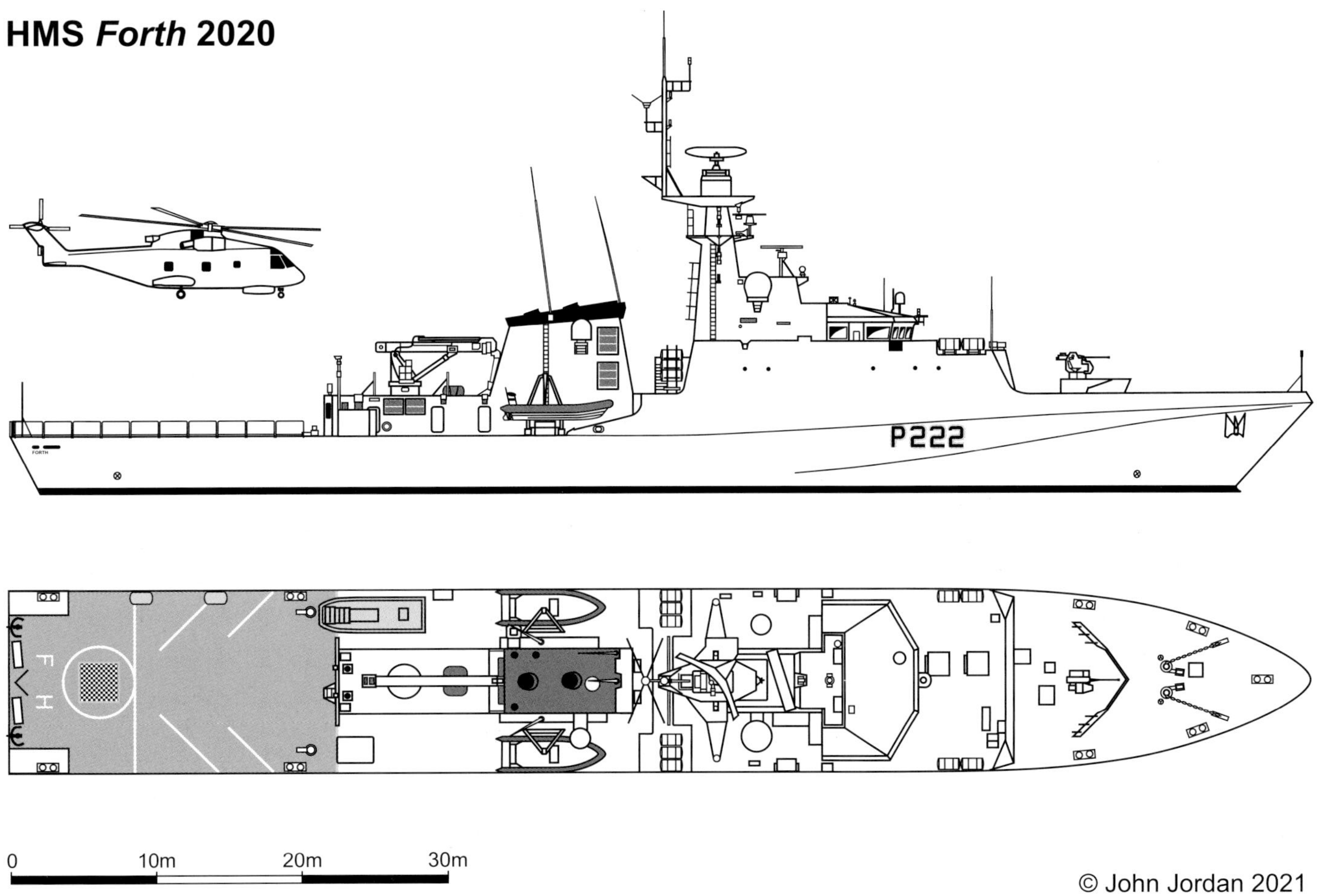

tonne handling capacity. This provides the ability to load and unload stores in locations where harbour facilities are limited.

Other Design Features: The Batch 2 'Rivers' are typically operated under a rotating three-watch system.[8] Under this arrangement, fifty (i.e. two watches) out of a total complement of seventy-five personnel are embarked at any one time. BAE Systems have, however, stated that the class can be operated by as few as thirty-four if required. Accommodation is provided for between sixty and seventy crew dependent on cabin configuration, all incorporating the high standards seen in recent Royal Navy vessels. There is austere, racked bunk space for an additional fifty personnel allowing, for example, the deployment of an embarked military force. This accommodation has a wide range of potential alternative uses, such as housing refugees in the course of a humanitarian operation or allowing trainees to be taken to sea.

The design is equipped with sufficient stowage space to conduct lengthy deployments, with autonomous endurance claimed as being up to 35 days. Independence of action is further supported by small but well-equipped workshops and a minor medical facility.

The two final Batch 2 'Rivers' incorporate a number of minor enhancements over the three previous vessels. These include changes required to comply with recent International Convention for

This detailed view of *Forth*'s forward superstructure shows shows some of the ship's key sensors. The large array for the Terma Scanter 4100 surveillance radar is positioned atop the mast, with SharpEye navigation arrays located further below, The Chess Dynamics Sea Eagle FCEO electro-optical director is mounted on a platform projecting forward from the superstructure. *(Conrad Waters)*

This view of *Medway* in the course of her Caribbean deployment shows two containers in the positions allocated for this purpose, with the starboard container surmounted by a large RHIB. Additional containers could be stowed on the flight deck if required. *(Crown Copyright 2020)*

the Prevention of Pollution from Ships (MARPOL) regulations, notably the fitting of catalytic converters to the exhausts to reduce nitrogen-based emissions. They are also equipped with an enhanced, wireless communication network to improve internal crew communications.

INTO SERVICE

When the plan to acquire an additional batch of OPVs was first announced in 2013, it was envisaged that the new ships would replace the first group of three 'River' class vessels on a one-for-one basis. These existing ships were allocated to the then Fisheries Protection Squadron and primarily tasked with policing waters off the United Kingdom.[9] However, the new batch was also seen as having the capacity for international deployment. The Batch 1 vessels *Severn* and *Mersey* had already demonstrated this potential during missions to the Caribbean in 2015 and 2016, whilst the modified Batch 1 'River' *Clyde* had been purposely built for operations off the Falkland Islands.

The subsequent decisions to acquire an additional pair of Batch 2 vessels and to retain the three Batch 1 'Rivers' in service to enhance the policing of British waters in the aftermath of the Brexit vote has seen the importance of the new vessels' international role come to the fore. The Batch 2 'River' class will play a prominent part in a new Royal Navy policy of international forward presence, particularly in lower

Tamar in dock whilst being prepared for her long-duration deployment to the Far East. She has been painted in a new camouflage scheme officially stated as being inspired by the 'Dazzle' schemes of the First World War but seemingly also having much in common with the Admiralty Disruptive Schemes of the 1939–45 period. It was initially envisaged that the Batch 2 'Rivers' would largely be assigned to policing home waters in replacement for the first group of the series but changed thinking is seeing them allocated to the overseas presence role. *(Crown Copyright 2020)*

Medway exercising with a Merlin helicopter from the auxiliary RFA *Argus* whilst deployed off the British Virgin Islands. The Batch 2 'Rivers' will be increasingly visible around the world as part of the Royal Navy's forward presence strategy. *(Crown Copyright 2020)*

threat areas. As well as providing greater scope for defence diplomacy, the new approach will reduce some of the pressure on the navy's overstretched force of surface combatants. The change in the OPVs' operational focus has been reflected in the renaming of the Fisheries Protection Squadron to the Overseas Patrol Squadron in 2020.

Forth was the first of the Batch 2 'Rivers' to take up a forward role when she arrived in the Falkland Islands in January 2020. Here she replaced the decommissioned *Clyde* as the Falkland Islands patrol vessel whilst also undertaking the Atlantic Patrol Tasking (South) role. The same month saw *Medway* depart British waters for the Caribbean, where she will undertake the standing Atlantic Patrol Tasking (North) for the next few years. *Trent* spent part of 2020 in the Mediterranean supporting NATO's Operation Sea Guardian and has subsequently been permanently deployed to Gibraltar, arriving at her new home base in April 2021. As already mentioned, *Tamar* and *Spey* will also soon depart overseas, being destined for extended deployment to the Far East later in 2021.

CONCLUSION

The Royal Navy's acquisition of the 'River' Class Batch 2 OPVs occurred almost by accident, arising as an unplanned consequence of delays to commencing the Type 26 frigate programme and the consequent need to maintain industrial capacity. As such, the new vessels were initially sometimes seen as expensive white elephants, significantly overly-specified for domestic constabulary policing and in desperate need of a meaningful role.

It is certainly the case that the programme has proved to be expensive, eventually amounting to £646m (c. US$900m) or £129m (c. US$180m) per ship. This compares unfavourably with the reported US$230m capital cost of the three similar vessels (i.e. c. US$77m per ship) originally ordered by the Trinidad and Tobago Coast Guard as well as with the budget for other OPV programmes elsewhere. However, much of this cost was already committed to shipbuilding infrastructure, making the actual acquisition expense considerably more manageable.

Trent, the third member of the Batch 2 'River' class, pictured in June 2020. *Trent* spent part of 2020 in the Mediterranean supporting NATO's Operation Sea Guardian and has subsequently been deployed to a new long term base at Gibraltar to maintain a Royal Navy presence in this part of the world. *(Conrad Waters)*

Two views of a newly camouflaged *Tamar* in June 2021. Although ordered almost by accident, it has been fortuitous that the Batch 2 'Rivers' have arrived to meet an emerging requirement for the conduct of sustained operations overseas. According to a Freedom of Information Act answer, the colours used in *Tamar*'s external paint scheme are the standard Royal Navy light grey – Light Weather Work Grey (BSC381C 676) – along with Medium Sea Grey (BSC381C 637), a bespoke Dazzle Grey (International Paints' Interfine 629HS, HYE29A/A), Dark Weather Work Grey (BS381C 677) and Black (BS 4800 00-E-53). In addition, in common with other Royal Navy vessels, the ship's weather decks are painted in Extra Dark Sea Grey (BS381C 640). The port and starboard sides are identical. Please also refer to the back cover for a colour view of the ship. *(Conrad Waters)*

It should also be noted that the Royal Navy OPV 90 variants have benefitted from significant improvements, particularly with regard to survivability. Whilst not immediately apparent from a cursory glance, these add significantly to the ships' overall capabilities but come at a considerable price.[10]

Moreover, the turn of events since the class was first ordered has meant that the new OPVs are arriving at a fortuitous time. The extended endurance and flexibility to embark additional crew and containerised equipment inherent in the 'River' class Batch 2 design makes them particularly well-suited to the new global presence role, their lack of a helicopter sustainment capacity notwithstanding. Like the Royal Navy's colonial gunboats of old, the Batch 2 'Rivers' are arguably ships of their time.

Notes

1. These earlier 'River' class design iterations are described in the author's 'British-Built Offshore Patrol Vessels', *Seaforth World Naval Review 2012* (Barnsley: Seaforth Publishing, 2011), pp.108–27. The first three 'Rivers' – *Tyne*, *Mersey* and *Severn* – were purchased outright by the Royal Navy in 2012. *Clyde* was sold to the Royal Bahrain Naval Force in 2020 after leasing arrangements expired. She is now named *Al-Zubara*.

2. BAE Systems had acquired the Vosper Thornycroft shipbuilding business from VT Group in October 2009 after a short period when the two groups had operated their surface warship yards in alliance under the BVT Surface Fleet joint venture.

3. The Secretary of State for Defence (Mr. Philip Hammond), *Hansard*, HC Deb 6 November 2013 vol. 570 cc251-54.

4. *National Security Strategy and Strategic Security and Defence Review 2015 – Cm 9191* (London: HMSO, 2015), p.31.

5. It seems some final outfitting work on a number of the class was carried out at Govan, possibly due to congestion at the Scotstoun facility due to programme delays.

6. The MAN 16V 28/33D diesel has provided a popular option for the current generation of larger OPVs, also being used in the Royal Netherlands Navy's large oceanic *Holland* class patrol vessels and the Royal New Zealand Navy's *Otago* class.

7. The Kelvin Hughes I band SharpEye radar is also capable of helicopter control so it is not altogether clear whether this or the Terma Scanter 4100 is primarily used for this purpose.

8. This crewing model has required adjustment over the course of the pandemic.

9. Whilst the patrol task extends to the entirety of British waters, fishery protection off Scotland is devolved to the Scottish Government and undertaken by Marine Scotland.

10. There is some confusion over Batch 2 'River' class's ultimate cost. The £646m programme cost quoted is taken from a Royal Navy press release of January 2021, which states this figure is £44m under the original approved cost of £690m. These figures are higher than the figure of £635m previously quoted as the value of contracts allocated to BAE Systems, the difference presumably relating to government-furnished equipment and other overheads. The overall value of the three-ship Trinidad and Tobago acquisition was a little over US$310m but US$87m of this total comprised logistical support and training costs.

U.S. NAVY
LCAC 101

3.4 SIGNIFICANT SHIPS

THE US NAVY'S SHIP TO SHORE CONNECTOR

Replacing the Landing Craft Air Cushion

Author:
Richard Scott

The practical realisation of the hovercraft, or air cushion vehicle, is generally credited to the British engineer Sir Christopher Cockerell, who became the driving force behind the design and demonstration of the Saunders-Roe SRN-1 prototype in the late 1950s. While the Royal Navy would go on to experiment with hovercraft in a number of roles, and still retains a handful of small assault craft, it is the US Navy and the US Marine Corps that have gone on to become the most enthusiastic advocates of the air cushion vehicle in the military sphere. The US Navy's Landing Craft Air Cushion (LCAC) is a high-speed surface connector primarily designed to provide ship-to-shore movement of Marine Corps personnel, vehicles and cargo during amphibious assaults. The LCAC also offers utility for the transportation of medical aid and relief supplies onto otherwise inaccessible shorelines, avoiding the need for a prepared port area.

As such, the utility of the LCAC spans the full spectrum of military operations – from major combat operations, where the high-speed air cushion vehicle can transport troops and material from large amphibious platforms beyond the horizon to landing areas ashore, to counter-insurgency, theatre security cooperation, logistical support, regional engagement, non-combatant evacuation and humanitarian aid/disaster relief. Furthermore, it provides a unique ability to traverse sand, marsh, tundra, snow and ice.

While the current LCAC fleet has given excellent service, proving themselves time and again in operational theatres worldwide, the craft are now approaching their end-of-life. To maintain its high-speed landing craft capability, the US Navy is in the process of acquiring a replacement design in the shape of the Ship to Shore Connector (SSC). Although similar in appearance to its predecessor, and drawing on an evolutionary pedigree, the SSC – now designated the *LCAC-100* class – is in fact a totally different craft, redesigned with greater load-carrying capability, and improved command and control systems. The US Navy has a requirement for seventy-three of these second-generation *LCAC-100* craft, comprising one Test and Training vehicle and seventy-two operational assets.

Left: The first operational production Ship to Shore Connector, *LCAC-101*, is one of a planned series of seventy-three second-generation US Navy hovercraft intended to maintain the navy's high speed landing craft capabilities. *(Textron)*

LCAC HISTORY

The US Navy has now amassed more than three decades of operating experience with its LCAC fleet. Derived from the prototype JEFF B air cushion vehicle tested and trialled during the 1970s, the LCAC provides high-speed amphibious surface lift for a 60-ton payload (75 tons in overload) at speeds of more than 40 knots. Initial operational capability was achieved in 1986; a first deployment was completed on *Germantown* (LSD-42) in 1987.[1]

Roll-on/roll-off loading and disembarkation is accomplished over a 4.5m stern ramp and 8.7m bow ramp. Range at full payload is 200 nautical miles at 40 knots, or 300 nautical miles at 35 knots. The craft can operate in Sea State 4 in unloaded condition and Sea State 3 fully loaded.

LCACs can be transported in, and operated from, all US Navy amphibious ships equipped with a well deck. The big advantage of an air cushion vehicle as a surface connector is to be able to operate independently of tides, water depths, ice, underwater obstacles, sand banks and beach gradients. The US Navy suggests that this increases the percentage of shorelines worldwide suitable for landing up to 70 percent, compared to 17 percent with a traditional Landing Craft Utility vessel. Against this, the LCAC is – in common with other amphibious air cushion vehicles – very weight sensitive, and suffers in high winds and heavy sea states. Furthermore, these craft are required to operate in a severe environment of saltwater spray and sand, which inevitably takes a toll on structure and systems.

LCAC-60, attached to Assault Craft Unit 4, prepares to enter the well deck aboard the *Wasp* class amphibious assault ship *Iwo Jima* (LHD-7) on 9 October 2020. A total of ninety-one LCACs were delivered to the US Navy between 1984 and 2001 with an original expected service life of 20 years. *(US Navy)*

A picture of *LCAC-65* pictured during operations in the Eastern Pacific in October 2020. The LCAC has remained in service far longer than initially anticipated, resulting in a requirement for a major service life extension programme to keep the type serviceable. *(US Navy)*

A total of ninety-one LCACs were delivered to the US Navy between 1984 and 2001, with production shared between Textron Marine and Land Systems and Avondale Gulfport Marine; a further six LCACs were built by Textron for the Japan Maritime Self Defence Force. The US Navy currently has seventy-two of the original craft still in service. While LCACs were originally built for a 20-year service life, most of the fleet will now serve for 30 years. To enable craft to be run on for an extra decade, the Naval Sea Systems Command (NAVSEA) has overseen a major service life extension programme (SLEP) since 2001.

The SLEP embodiment, performed during scheduled maintenance availabilities, addressed five discrete areas of the LCAC's sea-frame and avionics, viz.:

- Buoyancy box refurbishment (to address corrosion issues and introduce hull upgrades).
- Rotating machinery refurbishment (to extended useful life and reduce maintenance requirements).
- Enhanced engines (the four original TF-40B gas turbines being replaced by Vericor Power Systems ETF-40B gas turbines incorporating Full Authority Digital Engine Control [FADEC], which offer higher power, reduced fuel consumption and less maintenance).
- A new deep skirt with less drag, improved obstacle clearance, and lower maintenance.
- The replacement of the existing command, control, communications, computer and navigation system (C4N) with a new open architecture C4N suite.

SSC REQUIREMENT

Study and analysis work to identify a long-term replacement for the LCAC began in 2005 (the programme began life as the Sea-base to Shore Connector and for a time became known as the Joint Maritime Assault Connector). An initial capabilities document for the SSC programme was approved by the Joint Requirements Oversight Council in October 2006. In November the following year, the decision was taken to transition the programme into a government-led preliminary design and contract design phase. After approval by the navy's Resources and Requirements Review Board in December 2007, the programme began preliminary design in April 2008.

The top-level SSC requirement identified a need

for a worldwide-capable craft offering improved surface lift performance, able to achieve high speed (over 35 knots), increased payload (the cargo deck strengthened to carry a 74-ton load), operate from beyond the horizon (25 nautical miles or greater) and maintain capability in conditions through Sea State 3. At the same time, it was recognised that the SSC should deliver significant benefits in terms of reliability, maintainability and availability, with a commensurate reduction in lifecycle cost.

One constant was the need to fit inside the same amphibious ship well decks as the LCAC. This conditioned that the SSC should fit in the same space envelope as its forebear: 28m (91.8ft) in overall length, 14.5m (47.8ft) in overall beam and 8.5m (28ft) in height.

DESIGN APPROACH

The SSC programme was the US Navy's first major naval acquisition programme in more than 15 years to be designed 'in-house' by the navy rather than by private industry. Design activity drew on inputs from a number of centres and commands. These included NAVSEA's Future Concepts and Surface Ship Design Directorate, the Naval Surface Warfare Center Carderock, Panama City and Dahlgren divisions, the Space and Naval Warfare Systems Command, Naval Undersea Warfare Center Keyport and the Naval Research Laboratory. Some external contractor design expertise was also utilised: naval architecture houses CDI Marine (Band Lavis Division) and Gibbs & Cox provided concept design support.

The acquisition effort was iterative. It began with a Set-Based Design process, under which the craft level requirements were functionally decomposed into discrete, system level functional requirements documents (FRDs). These FRDs formed the functional basis for identifying the 'trade spaces', and to start preliminary design work. This was then followed by a contract design period. This developed the agreed baseline design and formed the basis for seeking detailed design and construction tenders from industry. Under the Set-Based Design approach, the navy took responsibility for 'locking in' the major ship characteristics (increased payload, operation in more severe environmental conditions, improved maintainability and reliability, and optimised total ownership costs). This meant that industry, bidding competitively against a common and transparent baseline, could focus on 'design

Two LCACs pictured in the well deck of the *San Antonio* class amphibious transport dock *Portland* (LPD-26) in March 2021. The ship to shore connector had to be designed to the same footprint as the previous landing craft air cushion to fit inside the same amphibious shipping. *(US Navy)*

LCAC-101 seen at speed in September 2020. The SSC design was the first to be produced 'in house' by a US Navy team for more than 15 years, with detailed design and construction contracted to a Textron-led team only once the major craft characteristics had been locked in. The design approach drew heavily on lessons learned from legacy LCAC operation. *(Textron)*

Two views of the prototype ship to shore connector, *LCAC-100*, in June 2020. She was ordered under a contract for the detailed design and construction of an initial SSC Test and Training Craft that was signed with Textron in July 2012 after evaluation of bids from two teams formed to compete for the award. The Textron team had the advantage of having built the vast majority of the previous LCACs and having gained additional experience in delivering LCAC upgrades. *(US Navy/Textron)*

for producibility' and reduced manufacture costs.

Building on lessons learned from LCAC operations, at the outset of the programme NAVSEA identified twenty-five reliability and maintainability drivers as priority areas to be addressed in the SSC design. Addressing specific design elements – such as the hull, machinery, auxiliaries and C4N – the SSC team then went through the process of developing a series of recommended design improvements. Among these were a number of material fixes intended to increase craft availability and lower life cycle maintenance costs. For example, it was decided that the SSC sea-frame – designed from the outset for a 30-year lifespan – would use more corrosion-resistant AA5083 aluminium and painted wet decks. It would also feature composite shafts and propulsor shrouds to reduce maintenance costs, and an improved skirt design to reduce drag and craft weight. Moreover, the machinery arrangement would be substantially revised, resulting in a simpler and more efficient drive train.

In March 2011, NAVSEA issued a draft Request for Proposals (RFP) outlining plans for a competitive solicitation for the detail design of the SSC, construction of a test and training craft (to be designated *LCAC-100*) for delivery in Fiscal Year 2018, and options for up to eight production craft, with a latest delivery date of May 2020. In 2011, Alion Science & Technology was contracted (under NAVSEA's SeaPort-e contract vehicle) to support the Amphibious Warfare Program Office (PMS 377) in the SSC acquisition effort.[2]

Two rival industry teams formed up to compete for SSC design and production: Textron Systems Marine and Land Systems came together with L-3 Communications (now L3Harris) and Alcoa Defense under the banner of Team Textron; while Boeing and Fincantieri-owned Marinette Marine formed Team SSC (which also included the participation of Griffon Hoverwork and Oceaneering International). Boeing later dropped out of Team SSC, with its role being taken on by Lockheed Martin. Following evaluation of the two rival bids, and a Milestone B acquisition approval in June 2012, NAVSEA selected the team led by Textron Systems to deliver the SSC programme.

Consequently, in July 2012, NAVSEA awarded Textron a US$212.7m fixed-priced incentive-fee contract covering the detail design and construction of an initial SSC Test and Training Craft (*LCAC-100*) for completion by February 2017. The

Ordered under a contract option exercised in 2014, *LCAC-101* is the first of eight low-rate initial production (LRIP) craft that followed-on from *LCAC-100*. The remaining LRIP craft were ordered between 2015 and 2016. *(Textron)*

contract included options for up to eight follow-on low-rate initial production (LRIP) craft which, if exercised, would bring the total value of the contract up to US$570m.

Team Textron had pitched its bid as inherently low risk, citing the extensive pedigree of Textron, Alcoa and L-3 in air cushion vehicle engineering, design, materials, systems integration, manufacturing, series production, and service life extension. Between them, these three companies could claim to have built no less than eighty-two new-construction LCACs – including the six supplied to the Japan Maritime Self Defence Force. They additionally brought experience of delivering LCAC SLEP upgrades.

Textron Systems also pointed to valuable corporate links with sister subsidiaries elsewhere in the wider Textron group. For example, Bell Helicopters would offer experience to SSC propulsion design and integration (specifically the complex integration of two gas turbine engines with cross-coupled transmissions using a graphite composite shaft) and the extensive use of composite components throughout its structure. In this regard, it is worth noting that the build and integration of the SSC is, in many areas, more akin to that of an aircraft than a warship. This includes the integration of modern avionics and fly-by-wire flight controls, as well as the use of high power density gas turbine engines in a small but lightweight package.

In December 2012 NAVSEA exercised a contract option to procure advanced planning, engineering and procurement services and long lead-time material for the first SSC LRIP craft (*LCAC-101*), the full contract option for *LCAC-101* subsequently being confirmed in August 2014. Later that year, in October 2014, NAVSEA announced the completion of the Production Readiness Review for the SSC programme. This milestone, together with a previously completed Critical Design Review, paved the way for the transition into production for *LCAC-100*. During these reviews, the navy closely evaluated SSC design maturity and readiness, the availability of materials and components, and industry's ability to successfully start and sustain manufacture.

Fabrication of *LCAC-100* formally began at Textron Systems' production facility just north-east of New Orleans on 17 November 2014. Work on the future *LCAC-101* commenced in January the following year. The options for the remaining LRIP craft were confirmed in April 2015 (*LCAC-102* and *LCAC-103*) and March 2016 (*LCAC-104* to *LCAC-108*).

Ship to Shore Connectors pictured at Textron Systems' production facility just north-east of New Orleans. This is the focal point for construction of all units of the type. *(Textron)*

DESIGN CHARACTERISTICS

While superficially similar to the existing LCAC, the SSC represents a substantially evolved design embodying new materials, systems and arrangements. For example, machinery design has been substantially simplified and rationalised: whereas the LCAC has two transversely-mounted engines per side, and a total of eight gearboxes (four per side), the SSC design features two longitudinally-mounted engines per side and just two gearboxes (one per side). As well as reducing power train gearbox complexity, this should also reduce maintenance overheads.

The engines themselves are more powerful and fuel-efficient Rolls-Royce MT7 marine gas turbines. Derived from the AE family of aero engines, and maintaining over 90 percent commonality with the AE1107C-Liberty (T406-AD-400) turboshaft powering the US Marine Corps' MV-22B Osprey tiltrotor, the 4-5 MW rated MT7 provides both propulsion and lift for the SSC. The twin-shaft axial design of the MT7 comprises a 14-stage compressor followed by an effusion-cooled annular combustor, a two-stage gas generator turbine and a two-stage power turbine. The engine is cold-end drive, featuring six stages of variable compressor vanes, a dual-channel FADEC system, modular construction and an 'on-condition' maintenance capability. Fuel and oil systems are fully integrated on the engine assembly. The MT7 shares a proven, common core architecture with the AE1107, AE2100 and AE3007 aero engines. Compared to the LCAC's ETF-40B gas turbine, the MT7 delivers approximately 25 percent more power while at the same time burning 11 percent less fuel.

Another key change concerns the lift fan arrangement. While the LCAC has two lift fans per side (with aluminium volute for lift fans and bow thrusters), the SSC design rationalises this to only one lift fan per side (with composite volute for lift fans and bow thrusters). Again, this revised design has been implemented with the intention of reducing power train complexity and lower maintenance requirements.

The SSC design features one auxiliary power unit (APU) and one gearbox-driven craft service generator (CSG) per side. This configuration will reduce energy costs in comparison to the legacy LCAC (which has one APU per side and no gearbox-driven CSG).

LCAC-101 comes onto dry land. This new generation of SSC incorporates significant technological improvements over its predecessors, including the use of a more durable grade of aluminium and the introduction of composite materials in a number of areas. These changes will all help ensure a longer service life. *(Textron)*

Composite technology has been introduced in a number of areas, notably the shafts and propeller shrouds (the LCAC uses steel shafting and aluminium alloy shrouds). Norwegian company Umoe Mandal has been given responsibility for the design and manufacture of a number of composite components. These include the all-composite propeller ducts, rudders and lift fan volutes with transition ducts, and air thruster nozzles.[3]

UK-based Dowty Propellers is responsible for the SSC propeller unit. Each craft uses two six-bladed controllable-pitch propellers. Dowty Propellers had previously supplied its R345 propeller system to operate as a shrouded airscrew on the current LCAC.

L3Harris Technologies has led on the development of an all-new C4N system designed to provide for a two-person cockpit in the starboard cabin (with redundant controls in the cockpit enabling single-pilot operation in an emergency). The company has built a System Integration Laboratory to support C4N engineering design and development for SSC software and hardware. A redesigned port-side cabin arrangement offers improved visibility for the loadmaster and deck engineers (who perform lookout functions while the craft is underway) and should also reduce fatigue.

One other change of note is the introduction of a 60 Hz power distribution bus; the LCAC has a 400 Hz electrical distribution system more common to aircraft. The objective is to provide for greater access to parts, and thus lower component costs overall.

PROGRAMME DELIVERY

The US Navy identified three SSC technologies as potential risk areas in terms of delivering the new craft: software; drive train components; and engine endurance. Additional testing was undertaken in an effort to further develop and reduce risk in these areas. However, the process of designing, building and testing a new naval platform typically throws up a number of engineering, manufacture and integration challenges, particularly when there is concurrency of development and production. The SSC programme has proved to be no exception, encountering a number of first-of-class technical issues. In particular, *LCAC-100* construction ran late, with builder's trials beginning in mid-2019 revealing a series of problems and deficiencies that required remediation. These issues created uncertainty and impacted the overall production schedule. Initial operational capability (IOC) – originally planned

Table 3.4.1
TECHNICAL COMPARISON – LANDING CRAFT AIR CUSHION AND SHIP TO SHORE CONNECTOR

TYPE	LANDING CRAFT AIR CUSHION (*LCAC-1*)	SHIP TO SHORE CONNECTOR (*LCAC-100*)
Delivered:	1984–2001	2020 onwards
Number:	91 (USN) + 6 (JMSDF)	73 planned (USN)
Design Life (Original):	20 years	30 years
Dimensions:	28.0m x 14.5m x 8.5m high Draught is 1.3m	28.0m x 14.5m x 8.5m high Draught is 1.4m
Designed Payload:	60.0 metric tonnes	74.0 metric tonnes
Propulsion:	4 x Lycoming/Allied Signal TF-40B gas turbines[1] 11.9MW (16,000 SHP) sustained	4 x Rolls-Royce MT7 gas turbines Over 18MW (24,000 SHP) sustained
Speed:	In excess of 40 knots loaded	In excess of 35 knots loaded (Sea State 3)[2]
Crew:	5	4

Notes
1. Replaced by 4 x Vericor Power Systems ETF-40B gas turbines rated at 14.2MW in LCACs subject to service life extension.
2. In practice, the actual speed of the *LCAC-100* design would seem to be higher than that stated given its higher propulsive output.

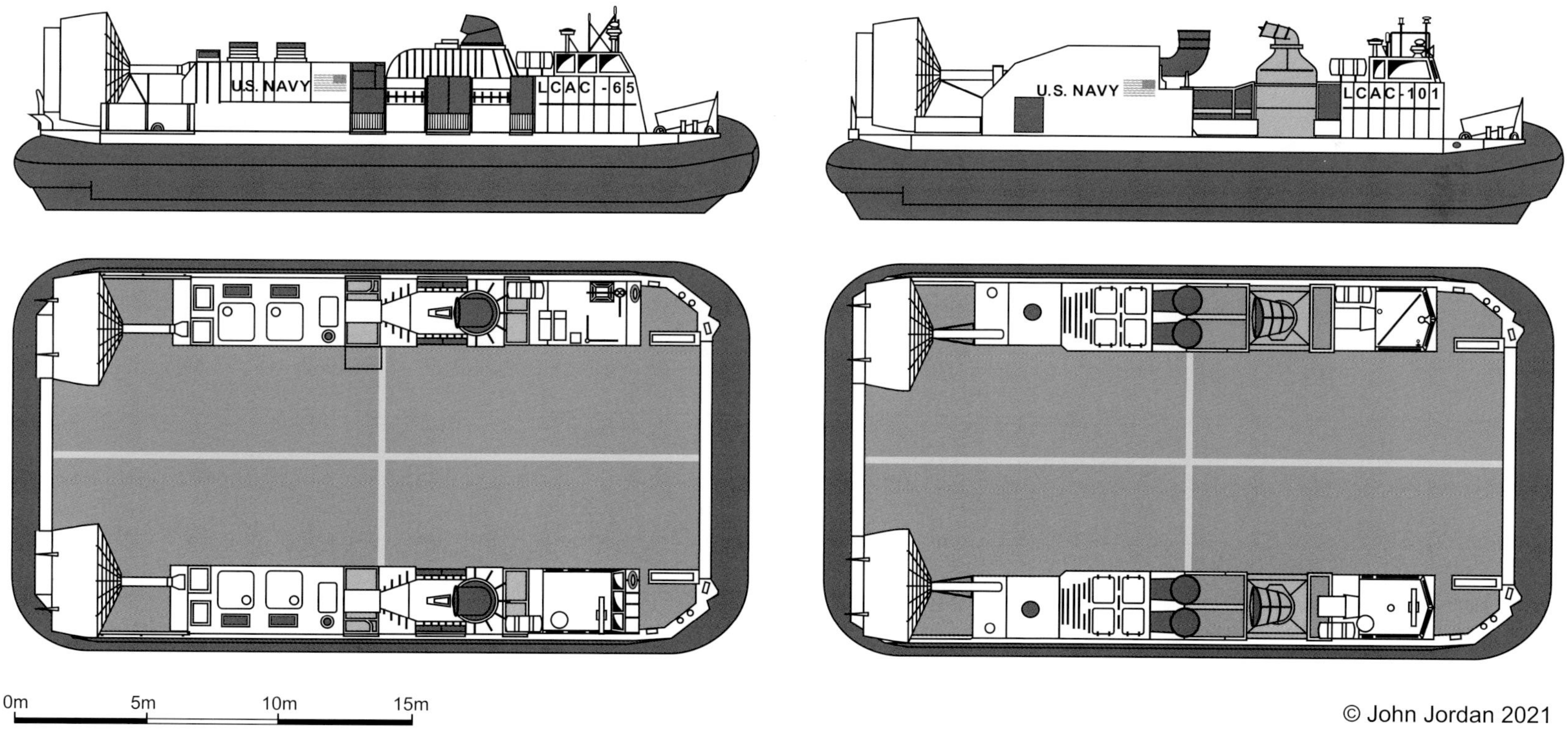

Above and opposite: *LCAC-100* and *LCAC-101*, the first two SSCs to be completed, seen arriving at the US Navy's Naval Surface Warfare Center Panama City Division on 2 September 2020. The Center's development, test and evaluation craft, *LCAC-91*, which escorted the two new vessels for the last stage of their journey, can also be seen. *(US Navy)*

for FY2020 – has been pushed back to FY2022 as a result.[4]

The June 2020 Government Accountability Office (GAO) *Defense Acquisitions Annual Assessment* threw light on a number of these problem areas.[5] For example, the GAO report revealed that the original gearbox design – part of both the propulsion and lift drive trains – had suffered various engineering issues requiring a number of redesigns. Also, pre-delivery testing showed that the rudder mechanism was not strong enough to direct the force of the propellers, and that the composite propeller blades themselves had suffered extensive cracking. Additionally, software development for SSC encountered delays. A forecast that 90 percent of code would be re-used from other programmes proved over-optimistic, and the engineering of the various software modules in isolation caused integration issues that subsequently required corrective action.

Textron modified *LCAC-100* to incorporate lessons learned during initial builder's trials. This resulted in a series of design remediations being incorporated into the craft prior to the completion of acceptance trials in December 2019. *LCAC-100* was formally delivered to NAVSEA in February 2020.

LCAC-101 – the first operational production unit – completed acceptance trials in early June 2020 after successfully completing a series of graded in-port and underway demonstrations for the navy's Board of Inspection and Survey. According to

NAVSEA, *LCAC-101* 'performed well having incorporated lessons learned from the recent [LCAC] 100 at-sea trials [and] successfully demonstrated the ability to operate both on and off cushion at full load through the full range of speed, payload and manoeuvring requirements'.[6] Formal delivery followed in late August.

'Fly-away' of *LCAC-100* and *LCAC-101* from Textron's New Orleans site was successfully achieved on 2 September 2020. The two craft transited to the Naval Surface Warfare Center Panama City Division (NSWC PCD), being met by *LCAC-91* for the last leg of their journey. Both craft are now operating in support of ongoing test and evaluation activities. *LCAC-100* will remain as a research, development, test and evaluation asset at NSWC PCD, being used to support continual development and integration of new technology and enhancements into the fleet.

SERIES PRODUCTION

With the SSC design now 'locked', lessons learned from the early trials programme have now been integrated into the production line. At the same time, Textron's investment in production process improvements and new plant is now yielding improvements in manufacturing efficiency and productivity: the company's New Orleans facility has adopted a streamlined manufacture and integration process, based on a highly efficient assembly station configuration, expediting outfitting at an earlier stage of construction and resulting in reductions in the number of hours it takes to build and test each subsequent craft.

Additional LRIP craft are now coming off the line. *LCAC-102* completed acceptance trials in April 2021 and was formally accepted by the US Navy on 3 June; *LCAC-103* is expected to begin trials in the third quarter of 2021. Follow-on orders for *LCAC-109* to *LCAC-123* (funded in FY2017, FY2018 and

LCAC-66 seen on the ramp at the Naval Surface Warfare Center Panama City in August 2019 just prior to being broken up. This particular craft had been used since December 1994 as an in-service special asset assigned to NSWC PCD to support research, development and test missions. The slower-than-expected delivery of new SSCs has meant that a LCAC Extended Service Life Extension Programme (E-SLEP) has been put in place to sustain the hovercraft fleet. *(US Navy)*

LCAC-100 and *LCAC-101* pictured in company. The new SSCs will sustain the US Navy's force of hovercraft-type landing craft for the next 30 years. *(US Navy)*

FY2019) were delayed while the technical issues affecting *LCAC-100* were resolved, and the design baseline stabilised. A contract for the production of these fifteen additional craft was eventually formalised by NAVSEA and Textron in April 2020. By May 2021, construction had commenced on craft through to *LCAC-116*.

To mitigate the slower-than-expected ramp-up in *LCAC-100* class numbers, the US Navy's FY2021 budget reallocated funding from the SSC programme to a LCAC Extended Service Life Extension Programme (E-SLEP) to sustain the connector fleet. Intended to cost-effectively squeeze more service life out of existing craft, the E-SLEP package is intended to address corrosion damage, replace obsolete electronics, upgrade the C4N suite (including cyber hardening) and replace the LCAC's deep skirt.

Up to eighteen LCACs will receive the E-SLEP embodiment, with all work to be performed at Assault Craft Unit Five at Camp Pendleton in California. In February 2021, Walashek Industrial & Marine was contracted by NAVSEA to undertake the first three upgrades, beginning with *LCAC-45*. The company's contract includes options for up to fifteen additional E-SLEP refits.

Notes

1. JEFF B – constructed by what was then Textron's Bell Aerospace – was one of two advanced development air cushion vehicle prototypes ordered for US Navy testing. The other was an alternative JEFF A design developed by Aerojet that was not ultimately selected for series production.

2. NAVSEA describes SeaPort-NxG as providing 'an efficient and effective means of contracting for professional support services and enhancing small business participation'.

3. Specialists in composite structures, Umoe Mandal were responsible for building the Royal Norwegian Navy's *Skjold* class fast attack craft. Amongst other significant projects, the company is also manufacturing composite mast structures for the Royal Navy's Type 26 frigates.

4. IOC will be achieved when six *LCAC-100* class craft have been accepted into service.

5. See, *Defense Acquisitions Annual Assessment GAO 20-439* (Washington DC: United States Government Accountability Office, 2020), pp.139–40 and available at: www.gao.gov/assets/gao-20-439.pdf

6. US Navy Team Ships Press Release, 16 June 2020.

4.1 TECHNOLOGICAL REVIEW

WORLD NAVAL AVIATION

A Review of Recent Developments

Author: David Hobbs

This year's overview of world naval aviation takes the re-emergence of the Royal Navy (RN) as a leading force in aircraft carrier operations as a major area of focus. Given also that various aircraft types frequently operate across a wide range of ships – particularly in the larger fleets – the chapter's layout has been revised from its traditional format to a structure based largely on national sub-headings. An exception is our concluding section, in which the MV-22B Osprey and its CMV-22B derivative are described in greater detail.

617 Squadron F-35Bs landing on *Queen Elizabeth* during the course of pre-CSG21deployment training. Additional F-35Bs will be acquired beyond the forty-eight already purchased but overall numbers remain uncertain. *(Crown Copyright 2021)*

UNITED KINGDOM – THE ROYAL NAVY

The United Kingdom's Integrated Review of security policy announced in March 2021 promised an enlarged fleet equipped with high quality, high-technology equipment across the spectrum of operations and confirmed that the two aircraft carriers are to be operated. Further batches of F-35B Lightning IIs are to be procured but probably in fewer numbers than originally planned. Prime Minister Boris Johnson spoke of safeguarding the United Kingdom's global influence and reinforcing the nation's ability 'to join the United States and other Allies to defend free and open societies'; aircraft carriers that focus power where and when it is required are central to this vision.[1]

CARRIER STRIKE GROUP 21 (CSG21)

On 4 January 2021 the British Government announced that *Queen Elizabeth* and her strike group had achieved initial operational capability (IOC). Full operating capability is planned for 2023 when 809 Naval Air Squadron (NAS), the second operational F-35B squadron, is commissioned. In early May 2021 *Queen Elizabeth* and her strike group took part in Exercise Strike Warrior off north-west Scotland with more than twenty warships, three submarines and 150 aircraft from eleven

British 617 Squadron and United States VMFA-211 F-35B Lightning II strike fighters pictured on the deck of *Queen Elizabeth* in May 2021 shortly before the commencement of the Royal Navy's CSG21 deployment. The integration of aircraft from two nations on the scale demonstrated by CSG21 is unique in modern times. *(Crown Copyright 2021)*

nations. It was the final work-up for her carrier strike group (CSG), with NATO's Standing Maritime Group 1 providing a most demanding 'enemy'. This concluded with CSG21, the task group that was to deploy to the Pacific, being certified as ready for deployment and operational command being passed from the RN Fleet Commander, Vice Admiral Jerry Kyd, to the Chief of Joint Operations, Vice Admiral Sir Ben Key.

In addition to *Queen Elizabeth*, CSG21 includes the Type 45 destroyers *Defender* and *Diamond*, the Type 23 anti-submarine frigates *Kent* and *Richmond*, an *Astute* class nuclear-powered attack submarine, the US Navy destroyer *The Sullivans* (DDG-68), the Dutch frigate *Evertsen* and the Royal Fleet Auxiliaries (RFAs) *Fort Victoria* and *Tidespring*. The carrier air group comprises eight F-35Bs of 617 Squadron Royal Air Force (RAF), ten F-35Bs of US Marine Corps Fighter Attack Squadron 211 (VMFA-211), seven Merlin HM 2s of 820 Naval Air Squadron (NAS), three of which are fitted with Crowsnest airborne early warning (AEW) role equipment, four Wildcat HMA 2s of 815 NAS and three Merlin HC 4s of 845 NAS.[2] Elements of 42 Commando Royal Marines are embarked to give a broad strike capability. With both RN/RAF and US Marine Corps (USMC) F-35Bs deployed, the air wing is the largest sea-based F-35B force to date and is the largest embarked on a British aircraft carrier since 1983. The Crowsnest-fitted Merlin helicopters are also the first to go to sea.

Some commentators have dismissed the embarkation of USMC F-35s as merely covering a shortfall in the number of British aircraft. However, in doing so, they have failed to understand the true significance of this integrated deployment. It is not unusual for an international group of warships to work together today, and for their aircraft, usually helicopters, to demonstrate a cross-deck capability in which they land, refuel and take off again from each other's ships. In contrast, integrated operations over a sustained period by embarked aircraft from two nations in the same air wing are far from easy. On the scale demonstrated by CSG21 they are unique. Before VMFA-211 could operate from *Queen Elizabeth*, all its pilots, maintenance personnel and support staff together with mission planning software, ready-use spare parts, nation-specific air weapons and American equipment that might not normally be found in a British aircraft carrier had to be taken on board and worked into daily operations. Both squadrons have then been required to work within a bi-national agreement that allows the free exchange of data to ensure that their aircraft are interchangeable within a common flying programme. Joint rules of engagement (ROE) have had to be agreed at a national level to allow the F-35Bs of either nation to operate to the full extent of their capability. In this instance a British carrier is providing global intelligence and, if necessary target data, to both British and American F-35B squadrons. Air support has to be made available, if necessary, for other coalition forces operating against a common enemy without causing embarrassment to either the United Kingdom or United States, with both countries anxious not to be drawn into conflicts that might be outside their normal ROE. If aircraft from both nations are combined in a strike mission tasked to use lethal, kinetic force, the decision on who has the right to authorise or abort it, on what grounds and at what level of command has had to be worked out well in advance.

The agreement covering CSG21 was signed by the British and US Defence Secretaries on 19 January 2021. It formally defined the terms of joint participation, encompassing both the long-term embarkation of VMFA-211 and also the attachment of the destroyer *The Sullivans* to the group. Ben Wallace, the British Defence Secretary, said that 'the UK now possesses a twenty-first century carrier strike capability, greatly assisted by the unswerving support and co-operation of the United States'. His US counterpart said that 'This deployment underscores the strength of our bilateral ties and demonstrates US-UK interoperability, both of which are key tenets of the US National Defense Strategy'. The fact that *Queen Elizabeth* can embark and operate another nation's strike fighters represents the culmination of decades of planning aimed at closer co-operation between the two navies to counter common threats. British taxpayers can take enormous pride from the fact that the RN has achieved something with *Queen Elizabeth* that no other navy has demonstrated; the long-term ability to embark an allied squadron during her first operational deployment. Ships from CSG21 are expected to visit more than forty countries and take part in over seventy pre-planned exercises. On the Mediterranean leg there has been a period of joint operation with France's *Charles de Gaulle* and the RN is to lead NATO exercises in addition to probable strike operations against ISIS targets in Syria. In the Pacific CSG21 will participate in Exercise Bersama Lima to mark the 50th anniversary of the Five Powers Defence Agreement between Malaysia, Singapore, Australia, New Zealand and the UK and throughout the deployment it is expected to work alongside naval and air forces from Australia, Canada, Denmark, Greece, Israel, India, Italy, Japan, New Zealand, Oman, the Republic of Korea, Turkey and the UAE as well as carrying out numerous diplomatic port visits. There could not be a better demonstration of the RN's restored global reach.

F-35B LIGHTNING II PROCUREMENT

Meanwhile, Jeremy Quin, Minister for Defence Procurement, has responded to Parliamentary questions by stating that the United Kingdom will eventually add to the forty-eight F-35Bs already procured. He also stated that the integration of the Meteor beyond-visual-range air-to-air missile into the F-35B is going well. Under current plans, Meteor-equipped F-35Bs could be deployed by 'the middle of this decade'. Initial funding for the integration of the SPEAR precision surface-attack missile with the F-35B is also already in place. The second operational UK F-35B unit, 809 NAS, is to form at the British F-35 base, RAF Marham, in the last quarter of 2021 and commission when it achieves IOC in 2023. Of the forty-eight F-35s procured by the UK to date, there will then be twelve in 809 NAS, twelve in 617 Squadron, twelve in 207 Squadron, the training unit, and three in 17 Squadron, the operational evaluation unit permanently based at Edwards Air Force Base in California. Although the unit cost of an F-35B has been reduced, the type's operating cost remains eye-wateringly high and this is reflected in the reduced numbers now being considered. The F-35B will be at the high end of a high-low force mix with affordable unmanned aerial vehicles (UAVs) providing mass and a broad range of capabilities against opponents with varying degrees of sophistication.

In October 2020 the US Navy's Chief of Naval Operations (CNO), Admiral Mike Gilday, and RN First Sea Lord, Admiral Tony Radakin, held a video conference hosted in *Queen Elizabeth* in which they announced their joint signature of a future integrated war fighting statement of intent that sets a co-operative vision for interchangeability. It is intended to synchronise the two navies' pioneering capabilities, strengthen operating concepts and focus collective efforts to deliver combined sea power together. Carrier strike is an immediate example of this co-operation and CSG21 may well provide answers about the optimal F-35 numbers for both navies. In March 2021, James Heappey, the British Armed Forces Minister, also briefed that the RN was studying other nations' plans to operate F-35Bs at sea with interest. In response to a question posed by the US Naval Institute, he said that the United Kingdom is 'open to mixing and matching its aircraft with other potential allies in the future', adding that 'there is going to be a community of F-35B nations that we would be mad to ignore'. He

Two views of the Crowsnest airborne early warning system being installed in a Merlin HM 2 helicopter. Declaration of initial operating capability for the system has been delayed by software issues associated with integrating Cerberus into the existing Merlin avionics but three Crowsnest-equipped helicopters are participating in the CSG21 deployment. *(Crown Copyright 2021)*

further stated that 'we think our carrier capability is something that we can develop alongside not just the US, but the Italians, the Japanese, the Australians and many others'.[3]

CROWSNEST AEW INTRODUCTION

In April 2021 Commander Chris Jones RN, the staff officer for Crowsnest Coherence within the Merlin HM 2 force, gave a briefing on the status of the project. He said that three pre-IOC standard aircraft were, by then, flying under RN control. The preceding months had seen RN aircrew and technicians preparing for the system's embarkation with CSG21. Crowsnest development had been delayed by savings measures and was incomplete when operations with the Sea King ASaC 7 helicopters that had previously performed the AEW role ended in 2018. Considerable extra resources had to be brought into the project to recover the lost time, which actually escalated cost beyond any savings achieved in the short term.

The Cerberus mission system at the heart of Crowsnest uses touch-screen technology for the two observers and an artificial intelligence (AI) based tracking system. It incorporates a library of data to identify airborne contacts which are fused with data from an integrated electronic support measures (ESM) system. Another feature is an enhanced mission recording and replay system which allows both real-time and post-mission data analysis and provides material for use in training simulations. However, development was hampered by software issues associated with integrating Cerberus into the existing Merlin avionics.

Once CSG21 had deployed only twenty-three Merlin HM 2s were available for continued Crowsnest development and training, anti-submarine training with 824 NAS, and embarkation in the second carrier – *Prince of Wales* – and in frigates. About one-third of this force is expected to be in maintenance at any one time and it is already apparent that the analysts who predicted that thirty was too small a force to carry out every planned task were right. There are a further twelve un-modernised Merlin airframes in storage but these have been extensively cannibalised for components, they are over 20 years old and their potential usefulness must be uncertain.

Despite appearing on the carrier-enabled power projection plan until 2026, the AEW-specialised 849 NAS was disbanded in April 2020 and its observers were absorbed into 820 NAS. Pilots for both Merlin HM 2 variants are drawn from a pool for either mission as necessary. The ten Crowsnest kits procured are to be rotated through the thirty Merlin HM 2 airframes and it was decided to absorb them all into the administrative structure of 820 NAS to plan the changes. However, the loss of 849 NAS makes little sense. AEW observer training still has to be given the priority it requires and anti-submarine training was deemed sufficiently important to retain 824 NAS as a separate training entity even though it, too, uses airframes from the HM 2 pool. It is not intended to change Merlin role-equipment on a short term basis but to make use of the most appropriate airframes as they are rotated through maintenance cycles. The appointment of a commander as a Crowsnest coherence officer indicates the importance of this system as a separate entity. The best way of retaining that coherence would be to retain a specialised naval air squadron to act as its focal point.

THE COMMANDO HELICOPTER FORCE

The Merlin HM 2s described above are administered by Navy Command. However, the RN administers a second Merlin force, the Commando Helicopter Force (CHF), under the operational control of the Joint Helicopter Command, JHC. In 2013 the decision was taken to transfer twenty-five Merlin HC 3/3As from the RAF to replace the CHF's long-serving Sea King HC 4. The HC 3 lacked blade-folding or lashing points and was, therefore, incapable of embarked operations. In 2014 the MOD therefore awarded a £455m (c. US$650m) contract to what is now Leonardo Helicopters to modify all twenty-five aircraft to a new HC 4 standard with upgraded aircraft management and mission computers, a tactical processor developed from the Wildcat design, and five 10in by 8in integrated touch-screen display units to control aircraft systems and mission equipment with two cursor control devices for tactical displays. In addition to the aircrew, tactical displays are used by the commandos in the rear cabin to maintain battlespace awareness before they are inserted in a landing zone (LZ). Communications are based on those in the HM 2 and there is an integrated defensive aids suite with additional chaff dispensers. Because the HC 4 will also be used in the CSG logistic support role, it has systems that allow it to fly in civilian-controlled airspace. These include twin GPS and a dual omni-directional range/instrument landing system. There is a new power-folding main rotor head and tail pylon, both pilot-controlled by cockpit switches. Lastly there was some strengthening to allow a fast roping and abseiling beam to be fitted to the cabin roof, allowing commandos to be roped down quickly when landing is impractical. Merlin HC 4s can carry an armoured Land Rover or a 105mm howitzer and ammunition internally or as underslung external loads.

By October 2020 the CHF had thirteen HC 4s serving with 845 and 846 NAS. They can be differentiated from HC 3s by their pale grey, rather than green, paint finish. Eight more were in Leonardo's Yeovil factory being upgraded with just four left to be taken in hand before programme completion in 2022. Replacement Merlin aircrew are trained for the CHF by 846 NAS which also provides aircraft for operational detachments and the maritime counter-terrorism (MCT) role.

The CHF also operates Wildcat AH 1s of 847 NAS in the battlefield reconnaissance helicopter role. The squadron's specialist role is the provision of control for joint fire from the sea in amphibious operations. It has gained an international reputation for its skill. Many of its aircrew are qualified as forward air controllers (airborne) (FAC(A)). They locate the enemy to provide intelligence, surveillance, target acquisition and reconnaissance (ISTAR) and can downlink video images and battle damage assessments. They are equally capable of directing naval gunfire onto enemy targets or directing fighters such as the F-35B. In 2016 a Wildcat from 847 NAS gave laser target designation to a USMC UH-1Y Venom, which used advanced precision kill weapon system (APKWS) rockets to destroy the target. In the same exercise another 847 NAS Wildcat carried out the first laser-spot target handover to a USMC F-35B.

The CHF embarks or deploys with its own combat support squadron which provides intelligence, engineering, logistical and medical specialists, vehicles, stores and specialised equipment including tentage. It includes a mobile air operations team (MAOT), which selects and controls LZs using satellite imagery and reconnaissance data from 847 NAS. As soon as an LZ is secured it directs aircraft movements, sets up fuelling points, and manages the movement of loads and the extraction of casualties. Merlin HC 4s can self-deploy, moving

A Merlin HC 4 helicopter from 845 NAS Commando Helicopter Force in flight over the British Virgin Islands in August 2020 during a deployment to the Caribbean aboard RFA *Argus*. The HC 4 variant of the Merlin has been modernised from the former HC 3/3A air frames transferred from the RAF to equip it for embarked operations. CSG21 has seen the type being given an additional role, that of combat search and rescue. *(Crown Copyright 2020)*

A Wildcat helicopter from 815 NAS is prepared for flight whilst operating from the Type 23 frigate *Lancaster* in the Baltic in March 2021. The Wildcat's abilities are being significantly upgraded with the entry into service of Martlet and Sea Venom air-to-surface missiles. *(Crown Copyright 2021)*

from RNAS Yeovilton to Cyprus in under 36 hours in one recent demonstration and consideration is being given to installing an air-to-air refuelling capability. Folded Merlins or Wildcats can also both be loaded into the hold of a Boeing C-17 for long-range deployment.

Merlin HC 4s have a new role within CSG21. This is joint personnel recovery, more commonly referred to as combat search and rescue (CSAR). Role training began with two pilots from 845 NAS qualifying as CSAR instructors in 2018 at the Gilze-Rijen air base in the Netherlands as part of a NATO course.The capability was part of the CSG IOC package declared at the start of 2021. The HC 4's rear loading ramp allows commandos to deploy and return rapidly during CSAR operations and three 7.62mm general purpose machine guns (GPMGs) mounted around the aircraft can provide suppressive fire backed up by snipers when a landing is opposed. The airframe has ballistic protection as well as the defensive aids suite mentioned above. An intelligence picture maintained through the helicopter's tactical processors and a maximum speed of 167 knots are both advantages in the role, as is the extremely low-level flight in which pilots are trained. The communications fit allows pilots to detect and home on their targets' personnel locator beacons, as well as linking with F-35Bs that provide top cover. The cockpits are night vision goggle (NVG) compatible for night operations.

The Royal Marines are also undergoing significant changes, typified by the formation of a new Vanguard Strike Company during 2020 to 'lead and inform how the Royal Marines will operate and fight in a dynamic and technological era of warfare'.[4] A Future Commando Force concept will see small detachments of CHF aircraft deployed as part of two Littoral Response Groups which will each comprise a few hundred commandos and supporting elements. One is to be based permanently east of Suez and the other is to focus on NATO's northern flank, working closely with the Norwegian armed forces. This unit would also be deployable rapidly to the Mediterranean at need. New technologies are being evaluated as part of this broader transformation, including small drones manufactured by Malloy Aeronautics of Maidenhead. These have been used for the autonomous transfer of various payloads of material using pre-designated GPS co-ordinates at ranges up to around 40 miles in a series of exercises.

WILDCAT HMA2 ENHANCEMENTS

The other aircraft type deploying with CSG21 is the Wildcat HMA2 operated by 815 NAS. Its air-to-surface armament has been limited to a door-mounted 0.5in machine gun. However, in 2021 the Martlet lightweight missile achieved IOC and the heavier Sea Venom missile is due to follow in 2022. The RN issued its requirements for these missiles, the future air-to-surface guided weapon (light) and (heavy) – FASGW(L) and FASGW(H) – in 2008 to replace the Sea Skua carried by its previous Lynx helicopters.

FASGW(L) is based on technology used by the Starstreak man-portable air-defence system developed by Thales at its Belfast facility. Martlet is fired from a hermetically-sealed launcher which has a shelf life of 15 years without maintenance. These are loaded onto a weapon wing in clusters of five, with one or two clusters on each wing giving a maximum of twenty weapons. Each launcher weighs c. 13kg, with a full load of twenty therefore approximating to 260kg. The wing has four hardpoints and generates 360kg of lift in flight to offset the weight of missiles but drag still limits aircraft performance to a degree. A Martlet accelerates rapidly to Mach 1.5 after launch and has a range of up to 8km. It uses a dual-effect shaped charge and pre-fragmented blast warhead and there are three selectable fusing options for use against specific targets: impact, height specific or delayed action. The missile is guided using semi-active control line of sight laser beam riding, which means that only one target at a time can be engaged. However, as each engagement only lasts less than four seconds from the missile being fired, defence against a swarm attack by hostile fast craft is feasible. Engagement begins with the observer placing a laser pointer on the target using a Wescam MX-15 electro-optical camera and selecting the auto target tracker. An active laser generation unit then transmits a low power beam from the Wescam turret and once the missile is in flight two receivers mounted on its tail detect the energy and maintain lock on target.

Contracts have been placed for all twenty-eight RN Wildcats to be fitted with mountings and wiring for the wings. However, when they are fixed in place the door-mounted machine gun – considered a more flexible weapon in some circumstances – cannot be fitted. It will only be possible to carry two of the larger Sea Venoms once these enter service as they weigh 120kg. It is possible to carry a mixed load of one Sea Venom and ten Martlets although, as yet, the RN has no stated requirement to do so.

FUTURE DEVELOPMENTS

Along with the Integrated Review, March 2021 also saw the announcement of exciting new projects for what the British Government referred to as the RN's future maritime aviation force. The Develop Directorate, part of the naval staff within Royal Navy Command at Whale Island in Portsmouth, is tasked with advancing future RN war fighting capabilities 'to achieve the optimum mix of present and future technologies for a modern, global and ready Royal Navy'. It wants UAVs to 'give wider options for the use of different types of air vehicle within the fleet'.

One of its proposals, already accepted by the Ministry of Defence, has led to a request for information from industry about assisted launch and recovery systems that could be fitted to *Queen Elizabeth* and *Prince of Wales* within the next three to five years. This should not be interpreted as a reversion to earlier efforts to equip one or both carriers with US Navy electromagnetic catapults (EMALS) and advanced arrester gear that were terminated in 2012. The request does specify an electromagnetic catapult but one with a maximum launch weight capacity of only 55,000lbs; 5,000lbs less than a fully loaded F-35B and 15,000lbs less than a F-35C. The maximum land-on weight for the arrester wires is 47,000lbs. Within these parameters a smaller fighter than the F-35B could be operated but the most likely reason for fitting assisted launch and recovery systems is to 'unlock' the carrier's big flight decks for a range of future, affordable fixed-wing UAV operations. Embarked F-35Bs would retain an important role in the opening phase of conflict against sophisticated opposition but UAVs would play an increasingly important part in routine operations against less sophisticated opponents and in so-called 'grey-zone' operations. They would be both cheaper to operate and more flexible, utilising the kind of technology now being demonstrated by Boeing Australia's Loyal Wingman UAV project which is now flying. The British carriers are expected to be in service until 2065 and are already beginning to show their

An AeroVironment Puma UAV operated by 700X NAS is made ready for launch during trials from the offshore patrol vessel *Tamar* in early 2021. The Royal Navy's Develop Directorate wants UAVs of different types to open up options for the use of aerial vehicles within the fleet. *(Crown Copyright 2021)*

potential beyond the present generation of aircraft.

The Develop Directorate wants UAVs to be part of the future maritime force 'by 2030', with Project Vixen standing out as the most exciting proposal. It is projected to be a medium-sized, fixed-wing, unmanned AEW aircraft intended to replace Merlin Crowsnest by 2030, thus freeing Merlin HM 2s for sea control and anti-submarine warfare missions. Additionally, Vixen could offer strike, electronic warfare, surveillance, air-to-air refuelling and data node specialisations with different role equipment packages fitted to a common airframe, adding mass and capability to the air wing without the need for constant flying to maintain pilot currency. It presents an exciting opportunity to use new technology to create a balance between manned and unmanned air systems to give greater combined effect and efficiency in the relatively near future.

In the meantime, much smaller AeroVironment Puma UAVs are being operated by 700X NAS which has three deployable units designated Phantom A, B and C Flights. Puma is a lightweight, hand-launched airframe with sophisticated cameras and a flight control system and each flight comprises air vehicles, a flight commander, an air engineering technician and a naval airman; all trained as remote pilots. It has a top speed of 45 knots and can stay airborne for two and a half hours tracking surface contacts and downloading high resolution images to its parent vessel. Relatively cheap – and easy to launch and recover – Puma is seen as significantly enhancing the intelligence-gathering abilities of ships that don't have other organic aviation capabilities embarked.

THE UNITED STATES NAVY AND US MARINE CORPS

The US Defence Budget for FY2022 released on 29 May 2021 saw the relative largesse of the Trump years come to an end, with Department of the Navy spending falling slightly in real terms. Although there was seemingly a particularly hard hit to aircraft procurement, with total spending on new aeroplanes declining by over 15 percent to US$16.5bn, Table 4.1.1 demonstrates that the numbers of planes acquired was largely in line with the old administration's plans. Notably, the previously scheduled end of F/A-18E/F Super Hornet and P-8A Poseidon production in FY2021 was confirmed after total runs of 678 and 128 airframes respectively. The planned withdrawal of legacy US Navy F/A-18A-D

Table 4.1.1: US NAVY PLANNED AIRCRAFT PROCUREMENT: FY2020–FY2022[1]

TYPE	MISSION	FY2020 Authorised[2]	FY2021 Authorised[2]	FY2022 Planned 2021[3]	FY2022 Requested 2022	CHANGE FY2022 Request-Previous Plan
Fixed Wing (Carrier-Based)						
F-35B Lightning II JSF	Strike Fighter (STOVL)	14 (10)	10 (10)	17	17	0
F-35C Lightning II JSF	Strike Fighter (CV)	20 (20)	26 (21)	20	20	0
FA-18E/F Super Hornet	Strike Fighter (CV)	24 (24)	24 (24)	0	0	0
E-2D Advanced Hawkeye	Surveillance/Control	6 (4)	5 (4)	5	5	0
Fixed Wing (Land-Based)						
P-8A Poseidon	Maritime Patrol	8 (6)	9 (0)	0	0	0
KC-130J Hercules	Tanker	3 (3)	5 (5)	5	6	+1
UC-12W Huron	Transport	1 (0)	0 (0)	0	0	0
E-6B Mercury	Airborne Command	0 (0)	1 (0)	0	0	0
F-5 Tiger II[4]	Training	22 (22)	0 (0)	0	0	0
Rotary Wing						
VH-92A	Presidential Transport	6 (6)	5 (5)	0	0	0
CH-53K King Stallion	Heavy-Lift	6 (6)	9 (7)	11	9	-2
C/MV-22B 22 Osprey	Transport	12 (10)	13 (9)	13	8	-5
TH-73A[4]	Training	32 (32)	36 (36)	36	36	0
Unmanned Aerial Vehicles						
MQ-4C Triton	Maritime Patrol	2(2)	1 (0)	0	0	0
MQ-9A Reaper	Surveillance/Strike	2 (0)	0 (0)	0	0	0
MALE-T	Surveillance/Command	0 (0)	0 (0)	0	6	+6
Totals:		**158 (145)**	**144 (121)**	**107**	**107**	**0**

Notes

1. The lack of FYDP (Future Years Defense Program) numbers in the FY2022 Presidential Budget Request makes it impossible to forecast procurement beyond that year.
2. FY2020 and FY2021 authorised numbers reflect numbers actually funded by Congress less some minor adjustments resulting from diversion of funding to pay for the Mexican border wall; figures in brackets reflect aircraft numbers initially set out in the Administration's Presidential Budget Request. As is frequently the case, the US Congress added material numbers of aircraft over the original request.
3. This figure refers to procurement plans set out in the FYDP figures presented with the FY2021 Presidential Budget Request, the last Trump Administration budget.
4. Refers to acquisition of surplus F-5 aircraft from Switzerland for refurbishment to act as 'aggressor squadron' aircraft in the combat training role.
5. Refers to the Leonardo AW119 variant selected to replace the TH-57 Sea Ranger (Bell 206 series) training helicopter.

series Hornets was also confirmed but is now being carried out to an accelerated timetable. More broadly, the budget was notable for its increased investment in longer-term research and development efforts as the navy attempts to reconfigure its forces towards a new focus on distributed maritime operations, particularly with a view to the emerging threat from China in the Pacific.

A glimpse of how US Navy aviation might evolve as an element within the distributed operations of a joint force was revealed in February 2021 when the *Dwight D. Eisenhower* (CVN-69) CSG commanded by Rear Admiral Scott Robertson took part in a sophisticated composite training unit exercise, COMPTUEX. This employed live, virtual and constructive (LVC) training tools. Admiral Robertson told the US Naval Institute that this was the first time that a CSG commander had taken charge of SEAL platoons, special boat detachments and marines conducting expeditionary advance base operations (EABO) focused on surface-to-air and surface-to-surface strike. The latter were inserted into the exercise virtually by the 2nd Marine Expeditionary Brigade's staff ashore. The exercise practiced maritime strike warfare, enhancing the CSG's lethality with tools such as over-the-horizon targeting, kinetic and cyber operations.[5]

Shortly after conclusion of the COMPTUEX, the *Eisenhower* group deployed for operational duties. It soon found itself participating in long-standing land attack operations against terrorist groups from the Mediterranean Sea, followed by deterrence against aggression from Iran after it passed through the Suez Canal and joined the Fifth Fleet area. However, the US is looking to reduce its presence in the Middle East to prioritise greater challenges elsewhere that might impact similar operations in future. In February 2021 the new US Defense Secretary, Lloyd Austin, ordered a global force posture review that will, inter alia, consider the need for a regular carrier presence in the Middle East as part of the increasing focus on the 'near peer threats' posed by China and Russia.

The possible procurement of light carriers (CVLs) for use in distributed operations is still an open question. Preliminary design studies are currently being undertaken by the Naval Sea Systems Command to help inform the ultimate decision. In February 2021, Rear Admiral Jason Lloyd, commander of SEA 05, explained that a cost engineering and industrial analysis team had been assessing various options to see what capability the navy might actually get from a CVL. Options ranged from an enhanced *America* (LHA-6) design to a '*Ford* class-light'. Previous CVL studies have also been re-examined because decisions made to reject them might no longer be relevant. Large nuclear-powered aircraft carriers clearly remain the USN's optimum design for operating manned aircraft but CVLs might, for example, be more suitable for operating a new generation of unmanned V/STOL aircraft. The integration of F-35Bs into amphibious assault ship air wings has already produced capable ships but more bespoke CVLs operating conventional carrier aircraft could bridge the capability gap between an amphibious ship and nuclear-powered carrier to create a wider network of strike fighters in distributed operations.[6]

The US Navy aircraft carrier *Dwight D. Eisenhower* (CVN-69) operating alongside the Japanese destroyer *Setogiri* (DD-156) in the Arabian Sea in April 2021. An American global force posture review announced by the incoming administration might make the presence of US Navy carriers in the Middle East a less frequent occurrence. *(US Navy)*

The second *Ford* class aircraft carrier *John F. Kennedy* (CVN-79) under construction at Huntington Ingalls Industries' Newport News facility in September 2020. Due for delivery in 2024, she was around three-quarters complete when this image was taken. Large nuclear-powered aircraft carriers such as *Kennedy* clearly remain the USN's optimum design for operating manned aircraft. *(Huntington Ingalls Industries)*

A F-35C Lighting II assigned to the 'Black Knights' from the Marine Fighter Attack Squadron 314 (VMFA-314) makes an arrested landing on the flight deck of the aircraft carrier *Abraham Lincoln* (CVN-72) in April 2021. The squadron achieved IOC for carrier operations in December 2020 and is to make its first overseas deployment aboard the carrier in late 2021. Their US Navy counterparts in the 'Argonauts' of VFA-147 will make the first operational deployment aboard *Carl Vinson* (CVN-70) earlier in the year. *(US Navy)*

CARRIERS AND AMPHIBIOUS SHIPS

The last twelve months have seen mixed news for the US Navy's force of aircraft carriers and large deck amphibious ships. Positively, the second *Ford* (CVN-78) class carrier, *John F. Kennedy* (CVN-79), will now be delivered under a single phase approach with her complete warfare system, including the ability to operate F-35C strike fighters, installed from the start. Both had previously been envisaged as post-build additions. The value of the contract to Huntington Ingalls Industries is up to US$315m but will save time and money in the longer term. The ship was reported as being 76 percent complete as of November 2020 and is scheduled for delivery in 2024.[7]

Ford herself made steady progress through her post-delivery test and trials phase, which ended in April 2021. Half that time was spent at sea and up to thirty-five aircraft were embarked. The next important milestone is first-of-class shock trials, which commenced in June 2021. After this, a maintenance and defect rectification period will be undertaken that will encompass certifying the final four of the eleven troubled advanced weapons elevators that are installed in the ship. The extent of rectification work that *Ford* will require is not yet specified but she is expected to carry out her first deployment in 2022.

Meanwhile, in July 2020 the amphibious assault ship *Bonhomme Richard* (LHD-6) suffered a devastating fire whilst alongside in San Diego naval base. A survey found that reconstruction would cost up to US$3.2bn, take five to seven years to complete and could only be carried out by delaying new construction. Disposal was the only real option and she was towed out of San Diego en route for the shipbreakers in Texas in April 2021. A valuable ship that had only just been upgraded to operate F-35s has been lost. *Bonhomme Richard*'s destruction coincided with the completion of the new amphibious assault ship *Tripoli* (LHA-7), which was commissioned at a low-key ceremony at Pascagoula, Mississippi after a planned public event had to be cancelled due to the pandemic.

FAST JETS

In December 2020 the USMC's first carrier-capable F-35C squadron, VMFA-314, demonstrated its ability to refuel and rearm aircraft at an EAB. Two aircraft landed within the USMC air/ground combat centre at Twentynine Palms, California onto metal matting and were arrested by an M31 expeditionary arresting gear system, which worked perfectly. Pilots remained in their cockpits with engines running as fuel was provided from a KC-130 tanker of VMGR-352 together with two 1,000lb GPS-guided bombs. After 15 minutes they took off to complete a scheduled bombing mission. VMFA-314 is to deploy with Carrier Air Wing Nine (CVW-9) in *Abraham Lincoln* (CVN-72) in late 2021 and the ability to hot refuel and rearm at an EAB is seen as being complementary to carrier operations; extending their reach when necessary. The squadron achieved IOC for carrier operations in December 2020.

Previously, in September 2020, an unspecified USMC F-35B unit at Marine Corps Air Station (MCAS) Yuma, Arizona, replaced the much-maligned Lockheed Martin autonomic logistics information system (ALIS) with its intended replacement, the operational data integrated network (ODIN). ALIS was originally designed over two decades ago and is now considered bulky and obsolescent. According to the F-35 Joint Program Office, ODIN immediately demonstrated improvement, even when compared with the latest ALIS software. ODIN hardware is a fraction of the size of the servers used to support ALIS, which weigh more than 800lb and comprise a man-high stack of electronics that need additional power modules that are difficult to deploy into austere environments. By contrast, ODIN servers comprise two transportable cases roughly the size of airline carry-on luggage, each of which weighs less than 70lb. ODIN processes information in half the time taken by ALIS and loads data from portable devices to the main server twice as fast, reducing the amount of time maintenance technicians have to wait for information to load. ODIN is also a cloud-native system with applications that can be regularly updated based on user feedback. After its initial roll-out in the autumn of 2020, one squadron at a time is being stood down and issued with hardware that supports both ALIS and ODIN. Squadrons will finally be divorced from ALIS one at a time from the autumn of 2021 but there is, as yet, no timescale for ODIN's adoption by the United Kingdom or other overseas operators of the three F-35 variants. ODIN's application to F-35s embarked in foreign vessels will therefore have to take into consideration the desirability of having all F-35 squadrons in the same ship using the same software.

The F-35 programme has come under renewed pressure from the House Armed Services Committee (HASC) Chairman Adam Smith in 2021 following a Government Accountability Office (GAO) report that the 2027 goal for Block 4 software IOC cannot be achieved. The cost of this upgrade rose by US$1.9bn between 2019 and 2020 and is now expected to exceed US$14.4bn. The GAO believes that software development is the problem, with a quarter of that delivered by Lockheed Martin found to have defects after its installation in aircraft. Continuing software issues have delayed the integration of upgrades to radar, electronic warfare, navigation, communication, weapons system and data fusion sub-systems, which may limit the type's combat effectiveness against sophisticated potential adversaries. The 2027 deadline was already a delay from the original 2024 target set for Block 4 by the Department of Defense in 2018.

Smith appears to be a strong advocate of reducing F-35 production numbers to release funds for less complicated fighters that could be operated more affordably. Significantly, this view is shared by General Charles Brown, Chief of the US Air Staff who has tacitly admitted that the original concept of the F-35 as an affordable, mass-produced fighter has been a failure. He too wants production numbers cut to fund an 'affordable fighter to replace hundreds of Cold War vintage F-16s', complementing a reduced fleet of sophisticated but costly F-35 Lightning IIs. After 25 years in development, the F-35 is now seen by many senior officers to have reproduced the problems it was originally intended to solve. Whilst admitted to be a vital asset in the opening phase of a high-intensity conflict, it is expensive to maintain, full of software bugs and complicated to deploy. Dan Ward, a former US Air Force programme manager, said that care must be taken not to use F-35s against unsophisticated opposition where its stealth and sensors are unnecessary.[8]

It will be interesting to see how this thinking impacts the eventual replacement of the F/A-18E/F Super Hornet inventory, which is now starting to become an important consideration. The latest Block III F/A-18E/F Super Hornet iteration has now reached the operational test and evaluation phase, with more coming off the line. However, older airframes will have to be retired from around 2030. As already mentioned, new F/A-18 production has been terminated with effect from the twenty-four ordered in FY2021, partly to make funds available for the next-generation air dominance (NGAD) programme. In late 2020 CNO Admiral Gilday briefed *USNI News* that the NGAD programme must avoid making the same mistakes that led to the F-35 taking over two decades to reach the fleet. Subsequently, in March 2021, Rear Admiral Gregory Harris, who leads the CNO's air warfare directorate, said that, under present thinking, the core aircraft that will follow the F/A-18 will 'most likely be manned' but that the programme will include a family of systems comprising both manned and unmanned options. As manned-unmanned teaming, exemplified by the Australian Boeing Loyal Wingman project, comes on line, this technology can be absorbed into NGAD.

NGAD studies are currently split into two increments. One focused on replacing the F/A-18E/F Super Hornet and the other the EA-18G Growler. It seems that the navy is ultimately driving towards an air wing in which 60 percent of the air vehicles will be unmanned but a lot depends on the success of the MQ-25A Stingray tanker. Notwithstanding the US Navy's obvious desire to get new technologies into service quickly, experience shows that doing so is far from easy and the next few years are likely to be crucial. The MQ-25A Stingray is one of the first major systems in the navy's emerging unmanned strategy. In July 2020 Boeing installed the aerial refuelling system into its T1 prototype to evaluate its aerodynamic properties and the software that controls it. The first carrier to be fitted with the unmanned carrier aviation mission control system (UMCS) is expected to be *George H. W. Bush* (CVN-

The US Navy's next-generation air dominance (NGAD) programme – destined to replace the Super Hornet – is likely to incorporate a mixture of manned and unmanned systems. The unmanned element could well draw on the artificial intelligence technologies being developed by Boeing under the Loyal Wingman programme for the Royal Australian Air Force. This is the first prototype at its official unveiling in May 2020. *(Boeing)*

77). The navy hopes to deploy the first MQ-25s in 2024 but has yet to determine a firm date.

US Marines from test squadron VMX-1 began flying the CH-53K King Stallion in January 2021. Previous developmental flight testing has been making good progress; this image shows the helicopter undertaking air-to-air refuelling trials whilst carrying an external load. *(Lockheed Martin)*

UNMANNED AIRCRAFT AND HELICOPTERS

The US Navy already operates unmanned air vehicles at sea in the form of MQ-8B and more recently MQ-8C Fire Scouts, mainly from Littoral Combat Ships. The 'B' models have been at sea since 2015, whilst IOC for the MQ-8C was declared in June 2019. They are operated by a total of two east and two west coast squadrons, with Helicopter Sea Combat Squadron 22 (HSC-22) on the east coast joining its west coast counterpart HSC-23 in operating the 'C' model in September 2020. The squadrons deploy mixed MH-60R and MQ-8 flights to Littoral Combat Ships, with pilots and maintenance technicians qualified on both types. The pilots fly the MQ-8s from a dedicated console in the ship's operations room. Fire Scouts have been successful in a growing number of mission profiles but the increased size and load-carrying ability of the 'C' version is offering new potential. In October 2020 Northrop Grumman carried out a navy-sanctioned trial with a launcher fitted to an MQ-8C capable of dispensing miniature sonobuoys manufactured by the British company Ultra. The idea is that the long endurance of the Fire Scout could extend the submarine detection range around the parent ship beyond that which could be achieved by an MH-60R. The search pattern could be up to 100 nautical miles from the ship. Not all MQ-8 operations have been successful, however. On 27 April 2021 a 'B' model of an HSC-21 flight embarked in *Charleston* (LCS-18) crashed into the side of the ship shortly after taking off and subsequently sank. The cause of the accident is not yet clear. No-one was injured and the ship was able to continue on her way to Guam.

Rear Admiral Harris has also spoken about the US Navy's future vertical lift maritime strike programme (FVL), which, like NGAD, is perceived as a family of manned and unmanned systems intended to replace the MH-60R/S helicopters from the late 2020s. The US Army and Marine Corps are also looking at FVL requirements but Harris has explained that the navy's need to operate from the decks of destroyers, frigates and Littoral Combat Ships are significant constraints that limit airframe weight, size and commonality. Common systems are possible, however and there is scope for working with the Marines' MUX unmanned aerial system, which also has to be compatible with surface combatant flight decks. Harris believes that experience with MQ-8C will help to inform FVL decisions. The US Navy recognises, like the RN, that there is scope for small VTOL UAVs to move logistics around a task force and Harris explained that many loads moved at present in manned helicopters

An MQ-4C Triton taxis at Andersen Air Force Base in Guam. US Navy MQ-4C Tritons of Unmanned Patrol Squadron 19 (VUP-19) have been operating from Guam for over a year. *(US Navy)*

weigh less than 40lb. Successful trials were carried in *Ford* during February 2021 using a small UAV to transfer packages.

US Marines from test squadron VMX-1 began flying the CH-53K King Stallion in January 2021. By the autumn, the Corps plans to form a four-helicopter detachment for initial operational test and evaluation. Once this is completed, IOC is anticipated to follow and the 'K' is expected to begin replacing legacy CH-53Es in 2023 or 2024. Training systems for both aircrew and maintainers at MCAS New River were certified as ready for use at the end of 2020 and, after training VMX-1, they will qualify personnel for the training squadron and then operational squadrons. At the smaller end of the scale, the navy and marines are looking at upgrades for the RQ-21 Blackjack UAV system. These include a vertical take-off and landing kit, a beyond-line-of-sight capability, a bandwidth-efficient common data link and a portable ground control station.

Meanwhile, MQ-4C Tritons of Unmanned Patrol Squadron 19 (VUP-19) have been operating from Guam for over a year. Whilst a small group of squadron personnel are forward-deployed to Guam, the mission operators are based at Naval Air Station Jacksonville in Florida with a permanent detachment of maintenance personnel at Naval Air Station Point Mugu in California, which is also expected to be the first shore base for the MQ-25A. Tritons have been integrated into Seventh Fleet operations and improved domain awareness across the Indo-Pacific region. The air vehicle can remain airborne for over 24 hours at over 50,000ft and has an operational range of 8,200 miles.

FRENCH NAVY

In December 2020, President Macron authorised the French Navy's *porte-avions nouvelle generation* (PANG), a nuclear-powered replacement for the existing *Charles de Gaulle* that is intended to enter operational service in 2038. Artists' impressions show a ship similar in configuration to an American nuclear-powered aircraft carrier and slightly larger than the British *Queen Elizabeth*. Information released to the press suggests she will have a length of c. 300m and a full load displacement of 75,000 tons. She will be powered by two K22 nuclear reactors and be equipped with electromagnetic catapults, advanced arresting gear and two aircraft lifts on the starboard side. Her air wing will be focused on around thirty SCAF next-generation fighters, which will be supplemented by AEW aircraft, combat search and rescue (CSAR) helicopters and UAVs. Development costs for the ship and its propulsion plant are estimated to amount to €900m (c. US$1.1bn) out to 2025 but likely construction costs have yet to be revealed. In January 2021, a further investment in French naval aviation was made when the French Navy announced that it is to replace its legacy Hawkeye AEW aircraft with three E-2D Advanced Hawkeyes. They will be delivered from 2028 onwards.

In similar fashion to the United Kingdom, France also deployed a CSG 'East of Suez' in 2021. In February *Charles de Gaulle* sailed for a four-month deployment to the Eastern Mediterranean, Indian Ocean and Persian Gulf designated Mission Clemenceau 21. The group undertook a series of exercises with allies and friendly powers before returning to Toulon in June. An amphibious task group led by the LHD-type amphibious assault ship *Tonerre* had left France slightly earlier in the month for the regular Jeanne d'Arc training deployment. Both groups followed similar routes and demonstrated deterrent presence, counter-terrorism and long-range strike capabilities. The carrier air wing comprised twenty Rafale M omni-role fighters, two E-2C Hawkeye AEW aircraft, two Dauphin CSAR helicopters and a single NH-90 Caiman helicopter. At the end of March 2021 *Charles de Gaulle* assumed tactical command of the US Central Command's Task Force 50, the second time she had undertaken this role. In April she undertook dual carrier operations with the *Eisenhower* CSG, reinforcing links between the two countries' naval air arms.[9]

ITALIAN NAVY

The aircraft carrier *Cavour* completed a sixteen month work package in Taranto Dockyard in May 2020 to prepare her for F-35B operations. This included adding a thermal-protective layer to the flight deck, improving data distribution networks, sensors and electronics and significant structural alterations to the hangar, island, briefing rooms and equipment stores. Fifteen F-35Bs have been authorised for Italian Navy service; two of these have been delivered and are being used for training Italian personnel at MCAS Beaufort in South Carolina. The Italian Air Force is to receive a further fifteen F-35Bs in addition to sixty F-35As; it would be fair to say they are being procured at the navy's expense since it has no plans to operate them from *Cavour*. The navy originally wanted

A computer-generated image of the French Navy's *porte-avions nouvelle generation* (PANG), which will replace the existing *Charles de Gaulle* towards the end of the 2030s. *(Naval Group)*

The first landings of F-35B Lightning II strike fighters onboard the Italian Navy aircraft carrier *Cavour* took place off the American coast on 1 March 2021 at the start of a period of tests and trials aimed at certifying the ship for the jet's operation. *(Italian Navy)*

twenty-two but numbers were reduced in 2012 as a cost savings measure.

Cavour arrived at Naval Station Norfolk, Virginia, in February 2021 to embark a USMC detachment tasked with allowing the ship to be certified for F-35B operation. Two specially instrumented US F-35Bs belonging to Air Test and Evaluation Squadron 23 (VX-23) were embarked for this purpose. All in all, the two F-35Bs carried out over fifty flight missions, some 120 vertical landings and as many short take-offs, with and without the use of the ship's ski jump. On satisfactory completion of the planned trials, *Cavour* subsequently embarked the two Italian F-35Bs to allow Italian pilots to conduct carrier qualifications. Although good progress has been made, *Cavour* is not expected to reach her IOC until the end of 2024 with the acquisition of eight F-35B aircraft and the training of twelve qualified pilots.[10]

The 33,000-ton amphibious assault ship *Trieste* is in the course of being fitted out as of mid-2021. She is equipped with a ski-jump on the forward part of her large flight deck and will also be equipped to embark F-35Bs, giving the Italian Navy more flexibility in deploying the type. However, it remains to be seen whether the navy will ultimately be allowed to procure enough aircraft to operate both ships with fast jets concurrently.

CHINA – PEOPLE'S LIBERATION ARMY NAVY (PLAN)

Reports from China suggest that the PLAN's third aircraft carrier, designated Type 003, is expected to be launched before the end of 2021 and enter operational service by 2025. It is speculated that she will have a full-load displacement of 80,000 tons and be equipped with up to four electromagnetic catapults of Chinese design. The two earlier carriers have been conducting joint operations, with a Chinese naval expert recently telling the state-run *Global Times* newspaper that 'China's two aircraft carriers will become key forces at a time when China has been facing military pressure from countries like the US in the Taiwan Straits and the South China Sea and potentially from India on China's key maritime transport lanes'.[11] The exercises are believed to have involved complex and high-intensity combat scenarios which analysts believe to have been designed to prepare the PLAN to counter a peer competitor threat attempting to interfere in what China openly refers to as the 're-unification' of Taiwan. The growing PLAN carrier force also looks to be deployed further afield. For example, in April 2021 it was revealed that a new pier installed at the Chinese naval base at Djibouti near the entrance to the Red Sea is big enough to accept aircraft carriers.

THE INDIAN NAVY

The 45,000-ton *Vikramaditya* was the only operational Indian carrier in 2021. The new indigenous aircraft carrier *Vikrant* completed preliminary basin trials at Cochin Shipyard towards the end of 2020 but had still to venture to sea as of mid-2021. Both are STOBAR ships with a 'ski-jump' and arrester wires but the Indian Navy believes that this system limits the types and performance of the aircraft it can operate and it wants a third carrier, the projected c. 65,000-ton *Vishal*, to provide the enhanced power projection capability it aims to achieve. Its procurement has seemingly been accepted in principle but opponents point to the navy's bitter experience procuring the existing carriers, which have suffered from significant delays and massive cost overruns. The new Chief of the Indian Defence Staff, General Bipin Rawat, has indicated that budget constraints will force the navy to defer plans for the new carrier, pointing to the desirability of assessing *Vikrant*'s operational performance and stressing the importance of other priorities, including submarines.

Since 2016 the Indian Navy has wanted to acquire new fighters to complement and, ultimately, replace its MiG-29Ks, which have apparently been a disappointment in service. Boeing has demonstrated F/A-18E/F launches from the shore-based ski-jump at Naval Air Station Patuxent River, Maryland, to show to the Indian Navy that the technique is viable from its two STOBAR carriers. Other options being considered include an improved MiG-29K design, the Rafale M and a navalised version of the Swedish Gripen. Although the Indian Navy has previously rejected plans for the immediate acquisition of a carrier-based version of the Indian-designed Tejas light fighter after deciding that it was overweight, its development has continued. In January 2020, a prototype undertook successful take-off and landing evolutions from *Vikramaditya*.

Boeing began delivering the first of a second batch of four P-8I Poseidon MPAs to India in November 2020. They will be operated at INAS Hansa in Dabolim in Goa as part of Western Naval

Boeing has demonstrated that the F/A-18 Super Hornet is capable of undertaking ski-jump launches in a programme of demonstrations held at Naval Air Station Patuxent River as part of efforts to sell the aircraft for operation on India's STOBAR carriers. *(Boeing)*

Command. The previous eight are operated by 312 NAS at INAS Rajali, Arakkonam, as part of Eastern Naval Command. American government approval for the sale of a third batch of six was announced at the end of April 2021. Meanwhile, in December 2020, Lockheed Martin rolled out the first Indian Navy MH-60R Seahawk, IN 751. The Indian government approved the purchase of twenty-one MH-60Rs in February 2020 and the first four aircraft were diverted from a US Navy production batch to enable early training for the Indian Navy after their modification to India's specification.

THE REPUBLIC OF KOREA NAVY

South Korea's 2021–2025 Defence Plan announced in August 2020 contained further details of the LPX-II project. In contrast to the existing amphibious assault ships of the *Dokdo* class, this will be a fully-fledged aircraft carrier intended to operate F-35Bs and helicopters and is increasingly being referred to as the CVX. Preliminary design has already been underway for some time and publicly released images show that the ship will have a twin-island configuration similar to that used in both the *Queen Elizabeth* class and Italy's *Trieste*. Britain's Babcock and Italy's Fincantieri are believed to have provided their input, the latter signing a contract with South Korea's Daewoo Shipbuilding & Marine Engineering (DSME) to support conceptual design in June 2020. Formal approval for the programme to proceed was given at the start of 2021, with delivery expected in the early 2030s.

THE ROYAL AUSTRALIAN NAVY (RAN)

Australian warships are to operate with CSG21 in the Pacific and, inevitably, the inclusion of VMFA-211 in the air group will stimulate discussion about Australian F-35B procurement. Even though no decision has been taken to procure STOVL Lightnings for operation from the RAN's ski-jump equipped *Canberra* and *Adelaide*, the fact that – by the late 2020s – F-35Bs will be operated by the UK, USA, Japan and South Korea in the Pacific must make the procurement of a batch of 'B' models attractive. If VMFA-211 can embark in the carrier of a close ally, so too could an Australian F-35B squadron. The ability of the Pacific allies to integrate assets in a way that has never been done before to achieve common capability must make sense, even to the most die-hard land-based air force lobbyist. They would achieve greater strategic effect than a larger number of F-35As flying localised sorties from an airfield in Australia.

A degree of rotary-wing interoperability has already been demonstrated by HMAS *Adelaide* in 2020, when she provided a deck for Republic of Singapore Air Force (RSAF) CH-47 heavy lift helicopter pilots to obtain deck landing qualifications. The RSAF CH-47 training squadron is based at Oakey in Queensland and already works closely with the Australian Defence Force. In another development, the RAN has short-listed five bidders for its SEA 129 Phase 5 rotary wing UAV requirement. This is intended to provide maritime unmanned intelligence, surveillance, reconnaissance and targeting to complement current ship-borne sensors and boost the navy's tactical surveillance capability.

THE V-22 OSPREY TILTROTOR FAMILY

The V-22 Osprey family of tiltrotors evolved from the Joint-Service Vertical Take Off/Landing Experimental (JVX) project that began in 1981. In 1983, Bell Helicopters and Boeing formed a partnership for the development and production of what became the prototype V-22A. Although initially a joint-service project, the USMC's need to replace the Boeing-Vertol CH-46 Sea Knight helicopter in

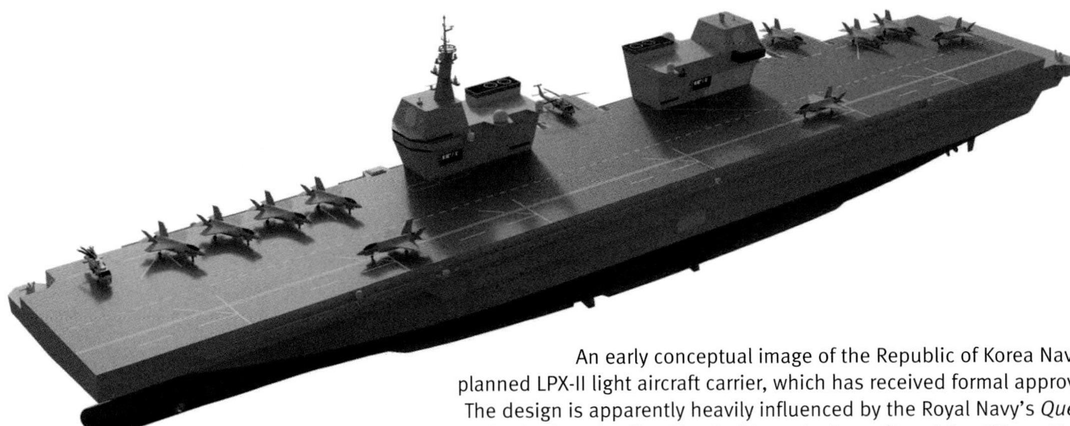

An early conceptual image of the Republic of Korea Navy's planned LPX-II light aircraft carrier, which has received formal approval. The design is apparently heavily influenced by the Royal Navy's *Queen Elizabeth* and Italian Navy's *Trieste* designs. *(Republic of Korea Navy)*

The first tiltrotor intended for large-scale military service, the MV-22B Osprey has been at the focal point of USMC operations since achieving IOC in 2007. This MV-22B – assigned to Marine Medium Tiltrotor Squadron 262 (VMM-262) – is pictured off Okinawa in May 2021. *(US Navy)*

service with their medium-lift squadrons saw the corps emerge as a main driver of the programme.

The resulting MV-22B for the USMC was the first tiltrotor intended for large-scale military service. A number of development problems had to be overcome along the way, which delayed the first V-22A Osprey flight until 1989. Operational evaluation began in 1999 and resulted in the V-22B redesign. This finally achieved IOC with the USMC until September 2007, when Marine Medium Tiltrotor Squadron 263 (VMM-263) deployed to Iraq.

The MV-22B subsequently carried out the first embarked deployment of the type as part of the 22nd Marine Expeditionary Unit (MEU) in *Bataan* (LHD-5) during 2009. Replacement of the CH-46 was completed in 2015 and MV-22Bs now form the core of every air combat element (ACE), within the corps' MEUs. As of 2021 the USMC has seventeen active-duty MV-22B squadrons and two Marine Reserve squadrons. A training squadron is based at MCAS New River, North Carolina and several units use the type in other roles. The USMC programme of record is a total of 360 aircraft.

The MV-22B is powered by two Rolls-Royce T406-AD-400 turboshaft/turboprop engines, each rated at 6,150hp (4.6MW) enabling vertical take-off at up to 47,500lbs. With a rolling take-off to combine wing and rotor lift the MV-22B can take off at weights up to 55,000lbs. For ferry flights – with an extra fuel tank fitted in the cabin – it can take off from airfields ashore at up to 60,500lbs, giving a ferry range of c. 2,100 nautical miles which can be further increased by in-flight refuelling. Procurement of the Cobham-designed V-22 aerial refuelling system (VARS) allows Ospreys to be used as tankers with up to 10,000lbs of give-away fuel to extend the range of F-35Bs or other MV-22s in an ACE.

A single 0.5in or 7.62mm machine gun can be mounted on the ramp but only covers the rear hemisphere; hardly ideal as aircraft approach an opposed LZ. Options for a forward-mounted weapon were considered before the Osprey entered service but were not developed until after the type had been delivered. The resulting BAE Systems defensive weapon system (DWS) comprises a gimballed 7.62mm Gatling-type minigun mounted through a hole in the floor of the rear troop/cargo compartment and controlled by a member of the aircrew with a console-mounted display. It has been deployed with several VMM squadrons but is not universally liked since the gun mechanism and its fire-control system take up space and weight and are not integrated with the aircraft's AN/ALQ-27 forward-looking infrared, FLIR. The non-flying pilot therefore has to 'talk' the gunner onto targets.

The MV-22B has brought game-changing capabilities; a single aircraft can carry twenty-four fully-equipped marines over a combat radius in excess of 400 nautical miles, even without in-flight refuelling.[13] It can hover like a helicopter or, when the rotors are lowered into the horizontal-flight position, fly as a turboprop aircraft at speeds of up to 305 knots at 15,000ft.[14] This performance has enhanced distributed operations, providing significantly more speed and range than legacy rotorcraft types and bypassing legacy landing craft with their limited radius of action and which had to fight their way towards a defended beach from the vulnerable, densely-packed landing ships that had carried them to the battle space. An Osprey carrier can remain hundreds of miles offshore in a widely dispersed group. It is difficult to comprehend why the third ship of the *America* (LHA-6) class is reverting to a well deck aft to operate landing craft at the expense of reduced air facilities.

The MV-22B's performance is leading to changes in USMC aerial assault operations, as AH-1Z Viper and UH-1Y Venom gunships lack the speed and radius of action to provide close escort and then strafe the LZ. A potential solution is the seizure of intermediate forward arming and refuelling points (FARPs). Remote FARP operations are practised to deliver an aviation-delivered ground refuelling (ADGR) capability. However, the USMC now uses fighters to escort MV-22B assault forces and these give faster response times, better overall offensive capability and a greater element of surprise. F-35Bs are especially effective since they also act as information nodes. Ospreys use satellite communications to enable long-range data reception, allowing marines to maintain situational awareness up to the moment when they leave the aircraft at the LZ, a capability known as digital interoperability (DI). Knowledge is always vital in warfare and the digital revolution is making it widely accessible down to individual pilots and marine section leaders.

Despite their robust construction and proven ability to survive in a combat environment, Ospreys have suffered from low serviceability rates for a variety of reasons, among them their large number of complicated moving parts such as wing and

This MV-22B Osprey was photographed aboard the Royal Navy carrier *Queen Elizabeth* in May 2021. The two Rolls-Royce T406-AD-400 turboshaft/turboprop engines, each rated at 6,150hp (4.6MW), enable vertical take-off at up to 47,500lbs total weight. *(US Navy)*

propeller blade fold mechanisms and their rotating engine nacelles. Original manufacturer's estimates of component life proved to be unduly optimistic, taking too little account of the impact of salt-laden sea air and desert sand. An element of redesign has eased the situation. However, continuing improvements have introduced another factor that complicates the type's operation; there are more than seventy different aircraft configurations across the fleet and both aircrew and maintainers have to know which they are operating. Savings measures have also resulted in spares deficiencies and cuts in depot level maintenance. As a result, serviceability has struggled to improve much above the 50–60 percent range. To overcome this, the USMC has introduced a V-22 readiness programme (VRP), which has included improvements in the supply of spare parts, better training for aircraft maintenance technicians and improved contract maintenance support. A key element of VRP is the common configuration readiness and modernisation programme (CC-RAM), which aims to update 129 Block B and 126 early Block C MV-22Bs to the latest Block C production configuration, reducing the number of different configurations across the fleet to five. The first MV-22Bs entered Boeing's modification line in 2018 and were re-delivered to the USMC in November 2019. The initial phase of CC-RAM is scheduled to be complete in 2022 but will likely be followed by subsequent technology insertion cycles.

The USMC's MV-22B has formed the basis of both the US Air Force Special Operations Command (AFSOC) CV-22B and, most recently, the US Navy CMV-22B carrier-on-board delivery (COD) variant.[15] The first CMV-22B arrived at Naval Air Station Patuxent River, Maryland, in February 2020 to begin test and evaluation with Air Test and Evaluation Squadron 21 (HX-21). The type was chosen to replace the Grumman C-2A Greyhound in the COD role in 2018 and has many advantages over the earlier type. Its passenger/cargo compartment is large enough to carry a replacement F-135 engine for embarked F-35C units, it has long range and its vertical landing and take-off capability mean that carrier catapults and arrester wires do not

US Marines are seen embarking into MV-22B Ospreys on the flight deck of the amphibious assault ship *Iwo Jima* (LHD-7) in this March 2021 photograph. The Osprey can carry twenty-four personnel over a combat radius of more than 400 nautical miles, transforming USMC operations. *(US Navy)*

have to be manned for COD operations when the ship is not at fixed-wing flying stations. It can also land on helicopter decks on ships other than aircraft carriers. Changes to the basic MV-22B design specified by the navy include increased fuel tankage to increase radius of action, improved cargo pallet handling arrangements, better lighting and a public address system for passenger briefing. There is space for 6,000lbs of freight or up to twenty-three passengers and an improved fuel-dumping system to get down to vertical landing weight. The CMV-22B can also be fitted with VARS to act as a tanker. It has a big part to play in the navy's distributed operation plans.

The US Navy initial programme of record called for forty-eight CMV-22Bs with the first thirty-nine added to an advance acquisition contract in July 2018. Since then four more have been contracted but it appears, as of 2021 that the navy intends to limit production to forty-four. Production is expected to end in 2025. The first operational unit to be equipped with Ospreys was Fleet Logistics Multi-Mission Squadron 30 (VRM-30) at Naval Air Station North Island, California, in June 2020. It is expected to achieve IOC in the course of 2021 and is to carry out its first deployment in the *Carl Vinson* (CVN-70) strike group before the end of the year. The first deck landings were carried out on *Vinson* in November 2020 as part of a trials programme made shorter by the type experience already gained by the USMC. In October 2020 a training squadron, VRM-50, was stood up at North Island and a second operational unit, VRM-40, is to be established on the east coast at NAS Norfolk, Virginia, in 2022.

The CMV-22B carrier-on-board delivery aircraft is the third Osprey variant to enter service. It made its maiden flight in January 2020 and will achieve IOC in 2021. *(Boeing)*

Notes

1. Prime Minister Johnson was speaking to the House of Commons on 19 November 2020 whilst announcing an increase in defence spending that predated publication of the Integrated Review's conclusions. See *Hansard*, HC Deb 19 November 2020 vol. 684 cc487-89.

2. Although formally a Royal Air Force unit, 617 Squadron is composed of both RAF and RN personnel, being currently commanded by a RN officer.

3. The United Kingdom's thinking was set out in an article by Sam LaGrone entitled 'MoD Official: U.K. "Would be Mad to Ignore" Working with Allied F-35B Capable Carriers' reported on the *USNI News* website – news.usni.org – on 23 March 2021.

4. Royal Navy press release dated 20 July 2020.

5. See Megan Eckstein, 'IKE Carrier Strike Group Commands SEALs, Marine Missile Teams in First-of-a-Kind, Large-Scale Drill' posted to the *USNI News* website on 17 February 2021.

6. See another Megan Eckstein article, 'Light Carrier Studies Already Underway As Navy Considers Role for CVLs in Future Fleet' reported on the *USNI News* website on 1 February 2021.

7. Huntington Ingalls Industries press release dated 2 November 2020.

8. See David Axe, 'The U.S. Air Force Just Admitted The F-35 Stealth Fighter Has Failed' posted to the *Forbes* website – forbes.com – on 23 February 2021.

9. The Clemenceau 21 and Jeanne d'Arc 21 missions are covered in more detail in articles by Xavier Vavasseur posted to the *Naval News* – navalnews.com – website.

10. For further information see Guy Toremans, 'The Italian Navy: Getting ready to operate the F-35B JSF' *Maritime Security & Defence* – June 2021 (Bonn: Mittler Report, 2021).

11. See Liu Xuanzun, 'China's two aircraft carriers in apparent simultaneous exercises for 1st time: reports' posted to the *Global Times* website – globaltimes.cn – on 6 September 2020.

12. The 2019 Marine Corps Aviation Plan calls for an eighteenth active MV-22B squadron to be formed but this had not yet occurred as of mid-2021.

13. In 2015 three aircraft of VMM-764 supported by four USMC KC-130T in-flight refuelling aircraft demonstrated the type's self-deployment capability by flying from MCAS Miramar, California, to Rio de Janeiro. They then participated in a rugged nine-day amphibious warfare exercise. In 2017 Ospreys of VMM-268 self-deployed from Marine Base Kaneohe Bay in Hawaii to Darwin in Australia, a distance of 6,000 miles using in-flight refuelling, albeit stops were made in Wake Island and Guam to allow brief relaxation periods.

14. Cruising speed is c. 270 knots at sea level whilst service ceiling is 25,000ft.

15. Overseas, Japan has also acquired the V-22 and a number of other countries, including Indonesia, have expressed an interest in the type.

4.2 TECHNOLOGICAL REVIEW

ELECTRO-OPTICS AT SEA

Some Considerations

Author:
Norman Friedman

For many decades, warships have been sprouting small electronic devices alongside their radars and radar countermeasures: electro-optical detectors of many kinds. Electro-optics is a generic term for the marriage of optics, which is far older, and electronics. Sometimes it is called optronics and occasionally photonics, the latter meaning a marriage of photons (particles of light) and electronics. The marriage means that something optical is converted into electronic form for further use. The earliest important example is television. An image is converted into electronic form so that it can be sent by radio, then received in a radio set and converted back into an image. In a closed-circuit television, the image may be transmitted only by wire. In recent years probably the most important development has been much more compact means of translating an image into digital form. That will be obvious to anyone using a digital camera or a camera embedded in a smartphone: the number of pixels the device can capture has increased enormously without requiring any increase in size. Because the image is captured in digital form, it can be processed digitally.

Compared to radar, an optronic sensor is passive: its operation is generally undetectable, which may be an enormous advantage. On the other hand, although it can track a target in bearing and in elevation, generally it cannot measure range, unless it is supplemented by some kind of rangefinder, often using a laser. Moreover optronics generally operates over much shorter ranges than radar. The exception is an airborne or spaceborne optronic sensor. This

Electro-optical devices are often used to supplement radar for a wide range of surveillance and tracking purposes. They can be particularly useful in littoral operations, where small targets can be lost to radar because of background clutter. This image shows two Norwegian *Skjold* class fast attack craft, which carry a Sagem VIGY 20 electro-optical tracker atop their masts. What appears to be an arrow pointing down at the horizontal, cylindrical VIGY 20 system is the ship's ESM sensor, which is located above a conical base that is intended to insulate it from the ship's emissions. *(Norwegian Armed Forces)*

Ghanaian Special Forces pictured operating from a small boat during a March 2021 exercise with the US Navy. Optronic sensors can often easily see small boats and the images captured can make it possible to decide whether or not they are hostile. Optronics can therefore provide a ship's command with the best possible means of gauging what and whether to engage in confused littoral situations. *(US Navy)*

Alleged Colombian drug traffickers caught in a frame from a forward-looking infra-red (FLIR) camera deployed by a US Navy helicopter. Imaging IR detectors first appeared operationally in the 1960s as airborne FLIRs. On board ships they are the basis for IR search and track systems. *(US Navy)*

discussion is limited to shipboard – including submarine – systems. Another key difference between a radar and an optronic sensor is that the latter generally captures much more detail, often making it possible to identify an entirely uncooperative target. With a radar, anything approaching such identification requires considerable sophistication.

Optics originally meant visible light, but typically the term now includes infrared (IR). In theory, IR is any radiation with a wavelength longer than that of light, which means that it shades over into the millimetre-waves used by some radars. However, for electro-optical purposes, what counts is the shorter-wave parts of the IR spectrum near the wavelength of light, at which radiation behaves much as light does. Typically radiation on the other side of the light spectrum (ultraviolet) has not been used, because it does not get very far in the atmosphere.

Because the wavelengths of visible light and usable IR radiation are so short, optronic devices tend to be far smaller than their radar equivalents. They offer far greater precision. In the important self-defence role, they can often detect an approaching missile before it clears the horizon, since they pick up its rising exhaust plume.[1] In a simulated scenario, for example, the plume was detected when the missile was 21.7km away and confirmed as a missile at 20.75km, but radar did not see it until it was 16.25km away. However, the radar measured the range to the incomer, and its speed could be measured by Doppler, so it could be engaged. A passive IR detector could not do that.

Because they are inherently small, optronic devices are often installed on board fast patrol boats and offshore patrol vessels in preference to radars. Also, thanks to short wavelengths, even a small device can track a target very precisely, since it is many wavelengths across. On the other hand, optical and IR ranges are inherently quite short. Weather limitations would seem to preclude abandoning even short-range radar altogether in preference for optronics.[2]

After the end of the Cold War, optronic detection had another perceived advantage. Navies found themselves operating more and more in littoral areas, confronted by small craft with very limited radar signatures. Such craft can easily be swallowed by radar sea returns (clutter). Optronic sensors often can easily see small boats at very short ranges. Moreover, in many cases the images captured make it possible to decide whether or not they are hostile. Classical forms of 'identification friend or foe' (IFF) are usually irrelevant, and small craft generally do not identify themselves using the automatic identification systems (AIS) applicable to ocean-going ships. Engaging a hostile boat may save lives, but attacking anything else may cause unacceptable civilian casualties. Optronics provides a ship's command with the best possible means of gauging what and whether to engage in confused littoral situations. That may not always be enough, but it is the best currently available.[3] For these reasons, force protection systems intended to protect a moored ship in port typically employ some form of optronics as their main sensors, the object being to detect and identify attacking individuals.

At least in theory, optronic surveillance devices are attractive to civilian ship operators faced by pirates and other criminal attackers. For example, in 2012 the German Diehl company announced an optronic shipborne surveillance system for merchant ships. Its sensors offered both day and night imagery, and it employed a degree of image recognition to reduce the workload of the attending crewman.

ELECTRO-OPTICAL DEVICES

Devices are often characterised by the wavelengths at which they operate, measured in microns (thousandths of a millimetre). Typical light wavelengths are about 0.4 to 1.1 microns (commonly 0.7 to 0.95 microns). Usable IR signals are limited by the 'windows' through which radiation easily passes, the most common being 3–5 and 8–14 microns. Radiation passing through these windows is sometimes called different IR colours. Typically the two common ones are called mid-wave IR (MWIR) and long-wave IR (LWIR). The hotter the source, the shorter the wavelength at which most of its radiation occurs. For example, a jet exhaust typically radiates most strongly at 3.75 microns, which is why many current naval devices operate at MWIR. Imagers intended to detect cooler objects operate best at LWIR, as was the case with the Diehl system referenced above. The shape of the curve of emission (strength vs wavelength) depends on the temperature of the source, so that a comparison between its radiation at two colours or even at two parts of the same colour ('shades') measures temperature. That can be a way of distinguishing a real object from a flare decoy.

Infrared – though not visible light – detectors generally have to be cooled to overcome heat-generated noise that would otherwise create spurious targets, although a few short-range hand-held devices need not be cooled. As detectors have shrunk, the cooling mechanism often accounts for much of the weight and mass of an optronic sensor.

Optronic detectors are typically divided into generations by the steady shrinkage of detectors. Generation I is a point detector, which may be scanned over a scene to create an image. The single spot detector in many IR-guided missiles is an example. So is the imaging tube in an analogue television. It focuses on one pixel of a scene at a time, registering the strength of the light falling on it. To create a full image, it scans across and rasters down between cross-scans. Resolution is generally measured in lines per frame – the number of rasters – resolution being the same horizontally and vertically.[4] Variation in the scene being scanned can interact with the scanning pattern to create interference. That is why, for example, people going on television were told to avoid striped shirts. Presumably a flashing light would have much the same effect on a television-based homing system.

As sensor technology developed, the single detector could be replaced by a cluster, each detector being too small to cover one dimension of a scene, but scanning more rapidly than Generation I. This is sometimes called Generation I+.

Generation II employs a linear array of detectors which covers one dimension of a scene. Digital cameras generally employ this approach, using a fast-moving mirror which reflects a slice of a scene onto the line of detectors. How quickly the scan can be carried out determines how quickly the camera can switch to a new image. Charge-coupled device (CCD) television cameras are in this category. Compared to Generation I, Generation II offers as much as a fourfold increase in resolution. The current Generation III employs large staring arrays, for example a 640 by 512 element. In the two decades after 1980, resolution more than doubled, and sensitivity – measured as a temperature difference – improved by a factor of about twenty; from the ability to distinguish a temperature difference of 0.2 degrees centigrade to an ability to detect 0.01 degrees. This rapid development was probably driven by the later stages of the Cold War; data on current systems suggests that progress over the last two decades has been relatively slow.

Imaging IR detectors first appeared operationally in the 1960s during the Vietnam War as airborne

An important function of naval IRSTs is to trigger a warning to a ship's defensive systems of an imminent missile attack. The French Sagem Vampir was developed to meet a NATO requirement for such a system in the early 1980s; its first installation was aboard the now retired air-defence destroyer *Cassard*, seen here in 2013. *(US Navy)*

forward-looking IRs (FLIRs). On board ships they are the basis for IR search and track systems. Search often means that the system reacts when a source of IR radiation suddenly pops up. It is usually up to the operator to decide whether it is significant. If it is, the system can be locked onto it. Analogue auto-tracking was initially used by aircraft. This technique was used by bombs such as Walleye, which were locked onto their targets before they were dropped. Later it was adopted to keep laser target designators locked onto targets, the bomb following the reflected spot from the airborne laser.

The technique was analogue, adapted to existing raster-scanned televisions. The television signal for one frame – as in a movie, which displays one frame after another – could be compared to that for the next frame. Any prominent part of that signal could be tracked. It has to correspond to a feature which remained recognisable as the bomb got closer or the aircraft manoeuvred. Typically it was either a well-defined edge or the centroid of a designated dark area. Much depended – and still depends – on whether the scene the tracker sees shows enough contrast or even the right kind of contrast. For example, an edge tracker can slide along an edge and then lose track altogether. It can be captured by an area of high-contrast onto which it slides. That might be the case, for example, for a ship and her wake. More sophisticated devices – presumably digital ones – were capable of correlation tracking, comparing successive views of a large part of a scene.

This type of operation might be compared to track-while-scan radar operation: the sensor keeps looking for new targets as it uses its data processor and memory to keep tracking one or more of them. The faster it scans, the better it can deal with rapid target manoeuvres, because the less the target image changes from scan to scan. Ultimately how rapidly the sensor can scan depends on how long it has to stare in one direction to form a usable image.[5] Users of digital cameras know how little time their sensors need to form a very elaborate image. Track-while-scan typically measures apparent target motion so that future target position can be predicted and compared to observed target position. Current digital IR search and track (IRST) systems can automatically detect and track sufficiently prominent targets, feeding their current and projected positions into digital combat systems.

During the Cold War, the important role of

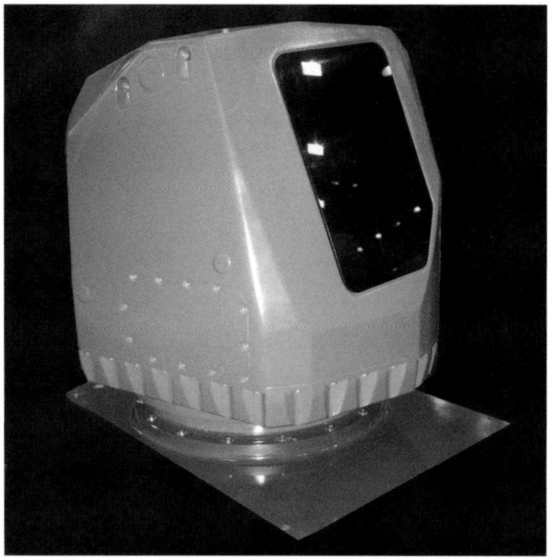

An image of the Safran EOMS NG naval electro-optical multifunction system, also pictured as installed on a United Arab Emirates Navy *Baynunah* class corvette. Drawing on a long legacy of French electro-optical technology, it combines long-range IRST with an EO director for gun fire control. *(Norman Friedman/Daniel Linares – Safran)*

Western naval IRSTs was missile warning, to trigger a ship's defensive system at short range. Development took considerable time. Although there were experimental naval IRSTs in the 1960s, NATO did not raise an operational requirement until the 1970s, and the French Vampir was chosen only in 1980. Its development seems to have been protracted; the first production version was installed on board the destroyer *Cassard* only in 1988. The delay may have been due in part to changes in how such systems were expected to operate. Initially Vampir was intended to auto-detect and track air threats. It was intended to operate in both IR 'colours', the operator seeing them as 360-degree swaths; the detector spun continuously. A third (processed) band highlighted targets the system detected; their positions went directly into the ship's combat system.

Vampir had to contend with the sheer mass of existing IR detectors and of the refrigeration needed to keep them cool enough. The solution was ingenious: the detectors occupied a fixed position under the spinning IRST sensor, a mirror that directed the radiation into them. One consequence was that the small Vampir turret could not carry other cameras. Many current IRSTs add a colour television (optical band) camera and a laser rangefinder. In the past, many turrets also carried a low-light-level TV camera. In many cases the operator has a joystick which enables him to designate a target for tracking.

The single IR eye of the IRST can scan in alternative patterns. For example, the current version of Vampir has two operating modes, horizon search (Ocean Mode, described as very long range) and the shorter-range Littoral Mode (elevation coverage). Effective range is shorter in Littoral Mode because the turret carries out an elevation scan while scanning horizontally, hence spends less time staring in any one direction.

A few search/track devices, such as the Israeli (Rafael) Sea Spotter, embody such massive arrays that they need not scan at all; they are equivalent to fixed-array radars. The resolution of its fixed arrays is limited, so the company offers an addition. An operator using the fixed arrays can cue a trainable turret – the Investigation Unit – which carries a conventional higher-resolution IR camera. Search time is limited because the turret is initially pointed at a spot near the target. The system uses staring arrays because it is primarily an anti-missile sensor. Staring arrays minimise the time to detect a fast-incoming missile; in its sales brochure, Rafael's illustration of the system emphasises the missile threat.

The tracking mechanism of a typical IRST can also be used to focus the device on a particular target, to create an electro-optic director. Such directors generally incorporate a laser rangefinder, since fire control is impossible without a measured range. Generally an IRST hands over a target to the director, which searches the area of uncertainty inherent in the director's track-while-scan operation.

IRSTs and IR fire-control systems are now quite common, though typically they attract much less attention than their radar counterparts. In some cases, radar directors incorporate IR or optical trackers as a back-up against jamming or to ensure tracking at low angles, where radar is unreliable because of multiple reflections.

Despite urgent interest in fast-reacting close-in defensive systems during the Cold War, not all navies found IRSTs sufficiently attractive. The US Navy seems to be the main case in point. Around mid-1972 the US Pacific Fleet became urgently concerned with what seems to have been the perceived threat of 'Styx' missiles in North Vietnamese hands. The fleet went so far as to place

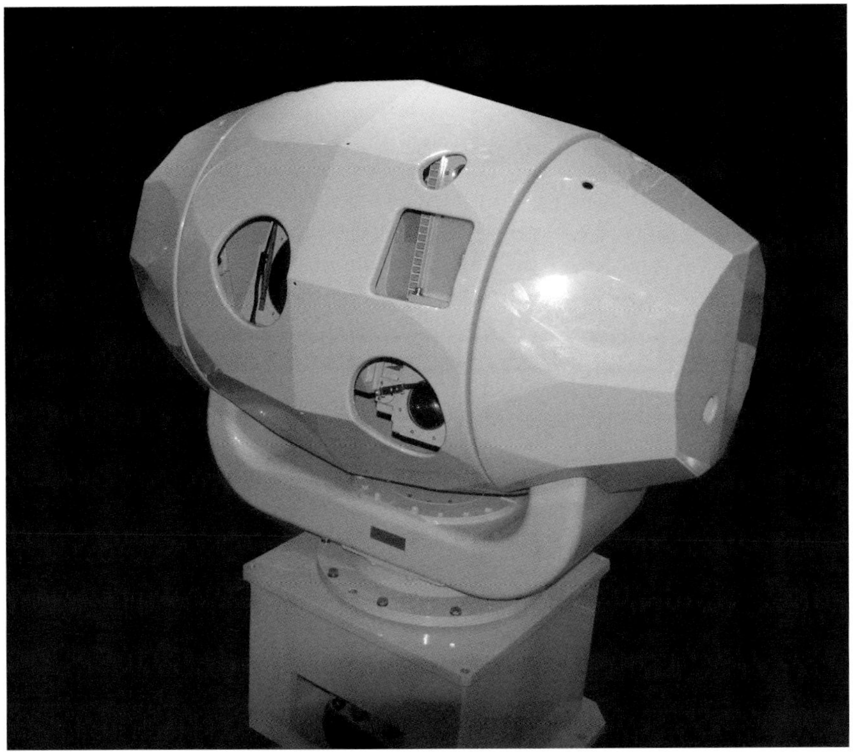

Although typically they attract much less attention than their radar counterparts, IRSTs and IR fire-control systems are now quite common. Pictured here is Ultra's Series 2500, two of which equip Royal Navy Type 45 destroyers (one on either side of the bridge wings). The windows cover a thermal (IR) imager (640 x 512 pixels), a CCD television camera (752 x 752 pixels), and a laser rangefinder. *(Norman Friedman/Conrad Waters)*

marines carrying Redeye anti-aircraft missiles in the catwalks of carrier flight decks. At this time the carriers' only air defence was typically the Sea Sparrow missile, which at that time used a manually-controlled radar director. NATO was developing a far more automated system to control the same missile, and the US Navy decided to adopt it. This system was probably the context for the NATO IRST requirement met by Vampir. The US Navy adopted a variety of pulse Doppler missile-detection radars, ultimately the Mk 23 Target Acquisition System and the current SPQ-9 radar.

The navy did also flirt with IRSTs. For example, the Target Acquisition System programme envisaged a Mk 24 iteration, which included an IR sensor. The navy also developed the dual-mode RAM missile, which relied heavily on IR and which developed into an all-IR fire-and-forget missile. This emergency effort was probably connected with a continuing US Navy IRST programme conducted with the Canadian SPAR Aerospace Company. In 1968–70 SPAR developed and tested an IR Electronic Warfare System (IREWS) for the US Navy. The name suggests that the emphasis was on decoying possible Soviet IR-guided anti-ship missiles; however, by the early 1970s the US Navy had concluded that nearly all Soviet anti-ship missiles were radar-guided. The Canadian Navy took over sponsorship. SPAR developed a Shipboard Passive Surveillance and Detection System, which became the subject of a Canadian-US project. The SPAR device was tested at sea in 1978–9, and a joint agreement for development was signed in 1983. However, the resulting SAR-8 was never adopted.[6] Nor were many later IR systems. The American view seems to have been that, even in littoral waters, its radars were good enough, and that very quick reaction to incoming missiles required the unique combination of detection and ranging which they alone could provide.

Most other navies do employ IRSTs and IR fire-control systems. Their light weight has made them particularly attractive for installation on board small combatants.

COUNTERMEASURES

As with radars, the major countermeasures against electro-optics are decoys, which for shipboard systems generally mean flares. For many years the issue has been whether signal processing or filtering would enable an IR detector to distinguish real from false targets. Filtering often exploits the dynamics of the situation: flares do not move the way a missile (or a ship) does. All optronic systems are also subject to jamming, for example by a flashing IR or light source which interferes with the scanning cycle inside the IRST. A staring IRST like Sea Spotter should, however, be immune to such deception.

There is also the IR equivalent of stealth. A missile designer can try to exploit the fact that outside the IR windows the atmosphere drastically reduces what a sensor can detect. Reportedly the IR colour of a missile exhaust can be shifted away from the two windows by doping. Stealthy jet aircraft do much the same thing by mixing cool air with their hot exhausts. It is not clear how well such measures work. The exhaust plume is probably just about all

A subsonic Teseo Mk 2A surface-to-surface missile firing from the Italian destroyer *Durand de la Penne*. IRST systems are an important means of detecting missile attack and are likely to become more so in a new age of hypersonic weapons. Unlike subsonic missiles, it will be difficult to hide their heat emissions by doping their exhausts. *(MBDA)*

The *Virginia* class submarine *Delaware* (SSN-791) pictured whilst running trials in October 2019. The use of optronic periscopes in submarines such as the *Virginia*s meant both that the command centre could be moved from its traditional position under the periscopes and that the imagery they provided could be made readily available to all system operators. *(Huntington Ingalls Industries)*

that an IRST can pick up from a very low-flying aircraft or missile.

A subsonic missile is probably not nearly hot enough to be very detectable by its body heat, although an IRST can generally see it against the sky. That would hardly be the case for a supersonic or hypersonic missile. IR is likely to become the principal detector of such weapons. Moreover, compared to a radar detector/tracker, an IRST gives no warning to an incoming missile it detects and hence will not – in itself – trigger crippling countermeasures.

OPTRONICS ON SUBMARINES

The most radical impact of electro-optics on warships has been on submarines. They already rely heavily on optics in the form of periscopes. The new technology has transformed the use of the periscope and its relationship to overall submarine design.

In the past, a periscope was an optical pipe connecting a lens and prism at one end with a prism and eyepiece at the other. The officer – typically the commander – using the periscope had to stand directly beneath the eyepiece, a requirement which profoundly affected submarine design. The top of the periscope inevitably created a visible disturbance in the water directly above the submarine. How long that lasted depended on how quickly the periscope observer could assimilate what he saw during a quick sweep. He was the only observer, and everything depended on how well he could identify what he saw in that sweep. Some periscopes were fitted with television or film cameras, but the quick sweep continued to define what a submarine commander had to do in order to understand a tactical situation.

France seems to have been the first country to adopt an alternative: an optronic periscope head feeding a computer. The head could scan very quickly – in one mode, a scan took 3 seconds – and whoever used the resulting image could take as long as necessary to interpret it. Very short scanning time compensated for the considerable size of the first French optronic periscope heads. Current heads are much smaller, and they continue to scan very rapidly.

The other significance of optronic periscopes is structural, on two levels. First, unlike the old periscopes, the scanning mast need not penetrate the submarine's pressure hull. Only wiring connecting sensors to the control room need do that. At the very

A graphic of Naval Group's SMX31 E concept submarine. One of the potential benefits of the use of optronic periscope heads that do not need to penetrate a submarine's hull is that the usual fin or sail could be dispensed with, reducing a major source of drag on underwater performance. *(Naval Group)*

least, that makes for a considerably stronger pressure hull.

The other dramatic change is that the control room no longer need be directly under the periscope. In the US Navy's *Virginia* class, for example, the control space is no longer in the cramped upper deck directly under the sail. Instead, it is on the widest deck in the ship, below that, and the periscope image is available to all of the system operators. That must be welcome. As submarines operate in littoral waters, their tactical situations become far more complex, and optical information becomes a mandatory means of making sense of that situation. Bringing the optical information into the control room in a form which can be examined at length makes it possible to integrate that data with everything else the submarine has available – acoustic information, ESM and intelligence data. It is no longer a quick means of assessing whether the ships near the submarine are threats or viable targets.

Breaking the mechanical connection between control room and periscope head suggests a more radical possibility. The optical sensor might be detached and placed on board a towed float. Sensors in the float might make it possible for the parent submarine to be sure of its position relative to that of the submarine. The float might carry other sensors as well. As long as it did not have to ride directly above the submarine, it would not give it away.

Moreover, without any need to project periscopes and other masts – except perhaps for a snorkel – a submarine might dispense with a fin (sail). The fin is a major source of drag, and its wake contributes to submarine noise via blade rate.[7] When the submarine turns, the fin causes snap-roll. At the 2018 Euronaval exhibition, the French firm Naval Group displayed a possible future fin-less submarine.

Notes

1. Typically an IR detector sees an approaching missile or aircraft as a dark spot against the heat of the exhaust plume. The plume may well be visible beyond the horizon, although it dissipates and cools as it rises.

2. Compared to radar, photonic devices are far more affected by weather and also by the evaporation layer over the sea, which can cause light and IR rays to bend (this layer also affects optical weapons: lasers). Longer-wave IR suffers less than light and shorter-wave IR in haze and fog. For example, a fog which halves transmission of visible light reduces medium-wave IR transmission by 13 percent but long-wave IR by only 3 percent.

3. This is why optronic systems are now often characterized as littoral or self-defence systems. During the Cold War, they were typically required more for close-in missile defence, as there was no perceived small-boat irregular warfare threat. Optronics, at least as now understood, cannot always be enough. There was no obvious indication that the boat which badly damaged *Cole* (DDG-67) in Aden in 2000 was filled with explosives. It is not clear, however, whether explosives have an IR signature.

4. For example, the standard US raster-scan format was 525 lines. The aspect ratio (horizontal vs vertical) was 4:3, so each line contained 700 picture elements (pixels: 4/3 the number of lines). The complete picture is a frame. As in a movie, the illusion of motion is maintained by changing frame quickly enough, in this case at 30 frames per second. Each frame consists of 367,500 pixels, which can be compared to the 12 megapixels – 12 million pixels – of many digital cameras. The US television standard was one of the earliest in the world, and foreign standards often offered higher resolution. For example, the British PAL used 625 lines.

5. Although the IRST may move smoothly, its scan can be imagined as a series of steps; what counts is how long each step has to take to achieve a desired level of vision, which means a desired ability to distinguish a target from the background. All scanning optical and optronic mechanisms can be seen this way. For example, how long a submarine commander has to take spinning a periscope to form a sense of the tactical situation depends on how quickly he can form usable images – which depends both on his eyesight and on the data processor in his head. For example, the stabilised turret of the Italian Leonardo Janus-N scans at a rate of about 86 degrees per second (it also scans vertically). Its Generation III IR sensor uses either a 384 x 288 or a 640 x 512 element array.

6. Attempts to integrate separate optical or optronic sensors with radars (i.e., as opposed to using such sensors as a back-up system) seem largely to have failed. For example, a low-light-level television sensor was mounted on the Sea Sparrow director, but the director slew proved too coarse for the television.

7. In the past, the US Navy was reluctant to dispense with submarine sails (fins) partly because the submarine still needs a bridge for use when entering or leaving port. Naval Group used a retractable bridge. It is not clear whether a package of towed sensors would be considered sufficiently reliable. The snorkel is currently needed even by nuclear submarines, since they have diesels for emergency propulsion in the event the reactor fails.

4.3 TECHNOLOGICAL REVIEW

Author:
Richard Scott

OPENING UP THE TOOLBOX

A New Model for Mine Countermeasures

It is recognised that the sea mine – laid in strategic chokepoints and channels, or in and around the approaches to ports and harbours – remains an enduring and potent asymmetric threat to both naval forces and commercial shipping. Mining has many attractions to those who would seek to deny others access in the water space. Mines are compact, easy to store in large numbers, have low maintenance requirements, and are easily deployed from naval vessels or non-specialist craft of opportunity. They are also relatively inexpensive to produce: the asymmetry of the mine derives from the fact that it can deliver an anti-access/area-denial effect wholly disproportionate to its cost. The mine is also unusual in that it presents a threat without any forces required to be present. A credible implied threat of mining may be sufficient to impact the maritime commander's risk calculus, and/or significantly increase insurance costs for commercial shipping. This psychological impact should not be downplayed: if the aim is simply to deter or delay, then no physical damage may be necessary for the mine to have achieved its mission.

British and US MCMVs pictured with the Royal Fleet Auxiliary *Cardigan Bay* in the course of the Artemis Trident 21 exercise in the Persian Gulf. The sea mine remains an enduring threat to both navies and commercial shipping, requiring a disproportionate effort to neutralise. *(US Army)*

Mines are both widely proliferated, and highly diverse in their nature. Old-fashioned 'horned' contact mines remain in abundant supply and, while relatively unsophisticated in technology terms, are still capable of causing extensive disruption to shipping. Bottom (buried or proud) mines, laid on the seabed and triggered by one or a number of influences – magnetic, acoustic, pressure, seismic and underwater electrical potential – represent an altogether more complex threat. Using advanced software-based control logic, these mines can select their preferred target according to a combination of influences, and can 'count' the number of vessels passing overhead before selecting a target. Moored mines can be laid much deeper than bottom mines and, apart from being actuated simply by contact, may also be triggered by influences. Some types can be tethered close to the seabed, receiving first passive and then active acoustic information before being released from their moorings and powering upwards towards their targets.

All the while, the mine threat continues to evolve.

The Royal Norwegian Navy minesweeper *Otra* encounters heavy weather during a NATO exercise. Such traditional MCMVs can be expensive to build and operate and feature significant limitations with respect to their ability to deploy at distance and to sustain a high operational tempo. *(Frederick A Bakken/Norwegian Armed Forces)*

The use of advanced casing materials combined with careful shaping has resulted in low echo strength mines that are increasingly difficult to distinguish from rocks and other debris on the seabed. There is also the emergence of a new genre of 'improvised' mines – the maritime equivalent of the improvised explosive device – to contend with. Unconventional in their design and manufacture, such devices have already been sown in waters off Yemen's Red Sea coast by Houthi separatists.

CURRENT MCM PRACTICE

The need for an efficient and effective specialist mine countermeasures (MCM) capability is well understood by navies worldwide, whether to 'punch a hole' for an amphibious assault, ensure access for seaborne logistics, to maintain freedom of navigation for mercantile traffic along commercial sea lanes, or to ensure the safety and security of ports, harbours and inshore waters. This means not only locating and neutralising mine threats, but also identifying those areas where mines are not actually present. The ultimate goal is to ensure the safe passage of shipping, whether by the provision of a 'swept' or cleared channel, or simply by re-routing vessels away from mined or suspected mined areas.

MCM operations in the modern era have, to an overwhelming extent, been performed by highly specialised – and commensurately expensive – mine countermeasures vessels (MCMVs) embodying characteristics of low signature and high shock resistance. The orthodoxy for most of the past half-century has been to send these manned MCMVs – hosting sophisticated sensors, remotely-operated vehicles and mine warfare tactical systems – into the mine danger zone to perform detection, classification, localisation, and neutralisation in slow and methodical sequence.

Search and classification operations are conducted using high-resolution sonar systems – hull-mounted, variable depth or fitted to a tethered self-propelled vehicle – that search the area ahead of the host MCMV. Once a mine-like object has been detected, a remotely-operated vehicle (ROV) will be deployed for identification and, if necessary, disposal. MCMVs may additionally embark a team of specialist clearance divers: these may be called upon to obtain threat intelligence, or perform ordnance disposal in conditions unsuitable for a remote control vehicle. A minehunter will typically have a crew of between thirty and fifty on board. These include the command team, sonar specialists operating the mine detection sonar equipment, inspection/disposal vehicle operators and clearance divers. Missions can last for just a few days, or may extend over several weeks.

Many navies also maintain a capability for minesweeping, a term historically used to describe the clearing of moored contact mines through the use of towed mechanical (wire) sweeps and cutters designed to sever mooring cables. As the nature and complexity of the threat has evolved, the definition of minesweeping has been expanded to cover all those devices and techniques used for clearing, actuating or jamming mines. Sweep operations are typically used in difficult environments where minehunting is judged to be impractical or ineffective (for example, shallow estuarine waters and sea areas characterised by rough bathymetry and high clutter density). While sweeps were historically towed behind a manned platform – placing the tow vessel at not-inconsiderable risk – more recent systems have employed remotely-controlled drones to deploy and deliver suitable influences.

However, a number of constraints and shortfalls are attendant to this classical approach to MCM. Dedicated MCMVs are expensive to build, maintain and operate, and they must venture into the mine danger area to execute their mission, slowing the operational tempo and necessarily increasing the risk to personnel and materiel. The 'platform-centric' nature of operations means the volume of search is necessarily limited to the number of hulls available. Furthermore, these hulls are slow to deploy and require significant support to sustain at range.

MCM operations are also punishing on human operators, who must concentrate for extended periods and apply their learned skills and experience to interpret the sonar picture. While modern sonar-processing techniques offer improved capability to locate low target echo-strength mines at range, and software tools can aid the operator with detection and classification, mine hunting remains a time-consuming, high-workload endeavour.

EMBRACING THE 'TOOLBOX'

As current MCMVs approach the end of their lives, and navies worldwide consider how they recapitalise their MCM capability, there is a growing shift away from 'platform-centric' forces in favour of a more modular 'toolbox' approach that embraces a range of systems, technologies and techniques adaptable to different platforms, environments or operational conditions. Central to this transformation is the use of unmanned and increasingly autonomous off-board systems – operating from less-complex mother ships, commercial vessels of opportunity, or shore sites – that can deploy sensors and effectors into the mine danger area at 'stand-off' ranges, so keeping personnel out of harm's way. Equally important is the accompanying introduction of smart software packages that enable MCM mission planning, autonomy, command and control, data processing and post-mission analysis.

'Taking the man out of the minefield' is an oft-repeated mantra, and there is no doubt that reducing risk to human operators is one major attraction of moving to a force design based on maritime autonomous systems. Yet this should not be seen as the sole driver for this new model of MCM operations: rather, the move 'off-board' – using unmanned vehicles that will swim, skim, or fly ahead into potential mined areas – is underpinned by wider analysis suggesting that autonomous systems are best placed to enable MCM to evolve to meet the developing underwater threats which are starting to impact traditional MCM techniques.

An MCM toolbox exploiting maritime autonomous systems breaks the tyranny that previously existed between the platform and the sensor, and in doing so offers an opportunity to expand search operations significantly through the deployment of multiple sensors into the mine danger area. This brings advantages with regard to operational tempo and survey quality (increasing area coverage rate through the use of unmanned vehicles equipped with high-resolution wide swath sonars), improved resilience in the face of unserviceability or attrition, component modularity and portability (enabling rapid deployment by land and air as well as ship), and the ability to rapidly reconfigure and scale the MCM mission package for specific tasks/threats (from a local route survey task up to a large-scale MCM campaign in support of chokepoint protection or an amphibious assault).

Another attraction of the toolbox approach is the increased flexibility its gives to capability managers and the acquisition community with regard to equipment delivery to the front line. Instead of a single 'monolithic' development and manufacture programme – with the attendant implications for cost, technical risk and programme schedule – a more incremental model can be adopted to enable acceleration of suitably mature technology to the user. This 'agile' mind-set also offers the ability to insert new capability rapidly so as to keep up with the accelerating pace of technological change.

Navies are increasingly moving away from conventional MCMVs to a modular 'toolbox' that relies heavily on unmanned and increasingly autonomous systems. Here, an unmanned US Navy RHIB deploying the AN/AQS-24 minehunting sonar system participates in MCM operations during an exercise also involving vessels from Oman and the United Kingdom in February 2020. *(Crown Copyright 2020)*

Of course, there are also risks and challenges associated with the delivering a robust and reliable off-board 'system of systems' solution integrating mission planning, detection, classification, identification, neutralisation and post-mission analysis. There are finely-balanced choices to be made on the composition, operation and support of the vehicles and systems in the toolbox, with alternative architectures conditioned or constrained by different operating concepts, mission profiles, force structure dependencies and budget provisions.

When it comes to systems engineering, there are many variables to consider: architecture and interface requirements (both functional and physical); vehicle launch and recovery in higher sea states; the degree of redundancy built into the system (including reversionary modes); interoperability between different systems; and the role/position of human operators within the detect-to-engage chain. Increased reliance on communications and 'over the air' data connectivity also imposes significant requirements around network capacity, resilience and security. This is made all the more complex if operations are to extend over the horizon.

It should also be remembered that the delivery of MCM effects is not just about equipment; the introduction of new and novel technologies embodied in the toolbox will necessarily have a profound impact on people, training, process, doctrine and infrastructure. For example, while there is an expectation that the move to maritime autonomous systems will lead to some reduction in manning levels, it is at the same time likely to drive a need for operators and maintainers with higher levels of technical proficiency.

UNDERWATER VEHICLES

One increasingly important component of the MCM toolbox is the free-swimming unmanned underwater vehicle (UUV), now increasingly referred to as an autonomous underwater vehicle (AUV) to reflect the increased level of autonomy embodied. Here, the naval community has largely sought to take advantage of technology already

A REMUS 100 man-portable UUV is recovered during a British Royal Navy exercise. Such systems are attractive thanks to their low logistics footprint and ease of launch and recovery. *(Crown Copyright 2016)*

US Navy sailors retrieve a Mark 18 Mod 2 Kingfish unmanned underwater vehicle (UUV) – based on the Hydroid REMUS 600 – during an exercise in the Pacific in August 2020. Medium-sized UUVs such as these can deploy larger payloads to greater depths and at greater ranges than their smaller counterparts but are more difficult to launch and recover. *(US Navy)*

proven in the commercial and research sectors, thereby accelerating delivery to the front line and enabling early 'hands on' experience. UUVs made their operational debut in 2003, when Hydroid REMUS 100 vehicles optimised for very shallow water survey were employed by the US Navy as part of port clearance operations in and around Umm Qasr in Iraq. Since then, a number of different types of UUV/AUV have been taken into service by MCM forces worldwide.

The small, man-portable UUV has proliferated most widely. Typified by systems such as the REMUS 100, General Dynamics Mission Systems' Bluefin-9, BAE Systems' Riptide-9, the L3Harris Iver3 and Teledyne's Gavia, this class of vehicle has found broad application for survey and explosive ordnance disposal (EOD) operations in shallow water and very shallow water environments. Such systems are attractive thanks to their low logistics footprint, ease of launch and recovery (being two-person deployable/recoverable), and their ability to accept a wide range of different payloads.

Medium UUVs have entered service in more limited numbers. Such vehicles, exemplified by General Dynamics Mission Systems' Bluefin-21 and the Hydroid REMUS 600, can deploy larger payloads to greater depths, and with extended endurance, making them ideally suited for mine reconnaissance operations where large areas of the seabed must be surveyed. However, their size and weight typically demands some sort of mechanised-assist launch and recovery system to enable the UUV to be deployed and retrieved over the stern or amidships of the host craft.

As regards sensor packages for survey and object detection, the primary technology employed by the military UUV user community is the side-scan sonar. As the name suggests, these systems scan a swath on either side of the host vehicle to capture an image of large areas of the seafloor, often co-registered with bathymetric soundings and sub-bottom profiler data. Commercial manufacturers are continuing to invest in affordable, low power-consumption side-scan sonar payloads delivering improved resolution and longer range. Accordingly, most navies use sensor payloads based on these same commercial off-the-shelf product lines, with many specifying dual-frequency systems to balance requirements for range and resolution.

However, the much-enhanced performance offered by synthetic aperture sonar (SAS) payloads has also attracted the attentions of the naval community. In essence, SAS overcomes the limitations of conventional side-scan sonars by exploiting data from multiple consecutive pings to synthesise a larger-array aperture. This means that SAS is not as limited by either range or resolution as traditional side-scan sonars, and is therefore capable of providing high-resolution images from UUVs. SAS is seen to offer a number of important advantages, such as high resolution at extended ranges, improved highlight data (to enhance classification performance) and increased search rates. However, its drawbacks are higher processing and power demands; a requirement for a highly stable and quiet host platform; and a unit cost significantly greater than that of a commercial off the shelf (COTS) side-scan sonar. There is also an enduring debate on the balance of quality versus quantity,

A Danish MCM flotilla comprising a minor standard vessel (MSF) in the foreground and a *Holm* class multi-role craft (MSD). The optionally-manned MSFs, which currently mount a Thales TSM 2054 side-scan sonar as part of the Royal Danish Navy's 'toolbox' of MCM systems, are being upgraded with Kraken Robotics' Katfish towed synthetic aperture sonar system. *(Guy Toremans)*

The Common Unmanned Surface Vessel (CUSV) developed by Textron Systems for the US Navy provides the basis for a multi-role Mine Countermeasures Unmanned Surface Vehicle (MCM USV) designed for operation from the Littoral Combat Ship and vessels of opportunity. *(Textron Systems)*

particularly the trade-off to be made with regard to attrition.

Some more bespoke sonar payloads have also been adopted. For example, the US Navy's Knifefish Surface MCM UUV marries the commercial Bluefin-21 deep water vehicle with a specialised Low Frequency Broadband (LFBB) sonar. The LFBB, originally developed by the Naval Research Laboratory in partnership with Ultra Electronics Ocean Systems and the Applied Research Laboratory of Penn State University, is optimised for buried mine detection. UUV/AUV technological development continues at pace. Specific focus areas include precision navigation, autonomy, sensors, communications, power and energy management, and propulsion.

UNMANNED SURFACE VEHICLES

The unmanned surface vehicle (USV) has an important part to play in the MCM toolbox thanks to its ability to perform a number of different roles. For example, it may serve as a 'taxi' by which to transport UUVs/AUVs into the mined area or swept channel; as a platform to tow a side-scan sonar or SAS payload (offering higher speed than an AUV); perform mine detection and classification using an integral hull-mounted sonar; be used to deploy tethered inspection/mine disposal vehicles; or to tow influence sweeps. Additionally, operation in an optional manned mode could allow the same platform to be used as a tender for diving operations.

USV design must reflect a number of different requirements. For example, the platform must offer space, weight, and power provisions to enable the installation and interchange of various different payloads and their associated handling systems; seakeeping characteristics should meet requirements for high speed in transit, and stability at very low speed for launch and recovery of unmanned payloads; a size/form factor that allows for portability and freighting; and a bridge/deckhouse space for optional manned operation. The ability to function with a high degree of autonomy is also of critical importance. Operation at range requires the vehicle, or system, to embed logic and reasoning so as to be able to build situational awareness, navigate in a safe, reliable and regulatory-compliant manner, and execute collision avoidance. This means that the USV only requires to be given a mission or 'goal' by the operator.

One example of an MCM USV is the Common

Unmanned Surface Vessel (CUSV) being introduced to service by the US Navy. Developed by Textron Systems, the CUSV is an 11m aluminium-hulled craft that provides the basis for a multi-role Mine Countermeasures Unmanned Surface Vehicle (MCM USV) designed for operation from the Littoral Combat Ship and vessels of opportunity. The craft features a retractable sensor mast, watertight electronics and forepeak compartments, and a large, reconfigurable payload bay area.

The first operational application for CUSV, as part of the MCM USV programme, is to deliver the Unmanned Influence Sweep System (UISS). Completing developmental test and operational assessment in November 2019, the UISS integrates a modified Mk 104 system acoustic generator and a magnetic minesweeping cable to provide for sweeping of magnetic, acoustic, and magnetic/acoustic combination mine types. Leveraging the same CUSV platform from the UISS programme of record, the MCM USV effort also aims to encompass additional payload delivery systems. These include a towed minehunting sonar system (either the AN/AQS-20C or the AN/AQS-24B) and a mine neutralisation capability (with the Barracuda unmanned mine-neutralisation system).

Atlas Elektronik has taken a similar multi-mission approach with its ARCIMS (Atlas Remote Capability Integrated Mission Suite) modular 11m USV. While early examples of ARCIMS were delivered for minesweeping operations, the craft has subsequently demonstrated integrations with a number of other payloads including towed side-scan sonars, UUVs/AUVs, and the Seafox expendable mine neutraliser.

UNMANNED AERIAL VEHICLES

Unmanned aerial vehicles (UAVs) can play two major roles as part of an MCM toolbox. In the first instance, the UAV can function as a communications relay between the MCM command operations centre – whether ashore or afloat – and the various unmanned/autonomous assets deployed ahead, thereby increasing range and enduring a robust and resilient network: communications generally rely upon line of sight (LOS) and there is a risk that a small vessel or vehicle operating in large waves will lose LOS visibility.

Having a UAV in 'orbit' allows UUVs/AUVs to surface periodically to exchange data packets via a radio link, meaning 'snippets' of sonar data can be

A British Royal Navy ARCIMS (Atlas Remote Capability Integrated Mission Suite) modular 11m USV pictured against a backdrop of conventional MCMVs whilst undertaking autonomous mine countermeasures trials on the River Clyde in 2019. *(Crown Copyright 2019)*

The MQ-8B Fire Scout UAV, currently serving aboard US Navy Littoral Combat Ships, has important MCM functions. Such UAVs can act as a communications relay between MCM command and the various unmanned/autonomous assets, as well as deploying sensors to locate floating or near-surface mines. *(US Navy)*

An image of the more capable MQ-8C Fire Scout UAV taken during long-range flight testing. The MQ-8C is destined to deploy the improved Coastal Battlefield Reconnaissance and Analysis (COBRA) Block II sensor package, which allows the rapid detection of surface and near-surface mine-like objects over a wide area. *(Northrop Grumman)*

relayed back to the command centre for processing while the mission is still in progress. Alternatively, it can provide a means to downlink an updated mission plan to deployed UUVs/AUVs.

A UAV operating at altitude over a potential mined area may also search for floating/near-surface mines if equipped with a suitable sensor payload. This is a course already being pursued by the US Navy with the AN/DVS-1 Coastal Battlefield Reconnaissance and Analysis (COBRA) Block I payload developed by Areté Associates. Integrated with the MQ-8B Fire Scout UAV, COBRA Block I uses a multispectral electro-optical/infrared sensor payload to conduct airborne tactical reconnaissance in the littoral battlespace for detection and localisation of minefields and obstacles in the surf zone and beach zone. Block I capability provides for the detection and location of unburied mine lines, minefields and obstacles on the beach in daylight. An improved COBRA Block II payload is planned for integration with the MQ-8C Fire Scout. COBRA Block II will increases capability beyond that of the Block I requirements to enhance the detection and localisation, in day and night, of surface and near-surface mine-like objects, (moored and drifting), mine fields, mine lines and obstacle detection and imagery collection from the surf zone to deep water at a high area rate.

COMMAND AND CONTROL

Smart software is an essential enabler for the MCM toolbox, and has a significant impact on the performance of the 'system of systems' as a whole. It is also an area where the move to an 'off-board' model, with multiple different robotic systems operating together at range, introduces additional complexity: while some navies may choose to acquire their toolboxes from a single vendor, others may elect to procure systems from multiple vendors. Whichever route is taken, there is necessarily a desire to select components that are open, interoperable and upgradeable to avoid vendor lock-in when replacing or upgrading toolbox components.

The requirement for comprehensive command and control system endures. As a key tool for the management of the MCM mission, the command and control system provides a software-enabled environment for mission planning, execution and supervision, assessment, post-mission analysis and training.

Sophisticated sonar image processing and display software is also playing an increasingly important role as a means to manage operator workload, increase confidence in the identification of mine-like objects, and reduce the possibility of human errors. These packages incorporate a range of facilities to aid detection, classification, contact management and reporting, including automatic target recognition techniques, together with specialist software tools for post-mission analysis. There is an expectation that developments in the field of artificial intelligence/machine learning techniques will enable the analysis process to be further automated and accelerated in the years ahead.

The planning and management of missions using multiple unmanned vehicles is one area where the role played by software has come increasingly under the spotlight. Most unmanned MCM systems come complete with basic autonomous capabilities that enable the preparation of mission plans for each individual system. However, trying to programme individual mission plans for multiple different assets, de-conflict these plans and then manage each system through the course of an operation would incur significant overheads with regard to time and manpower.

Some sort of open and modular operating framework is therefore required to ensure smooth and efficient integration and interoperability. For example, the UK's Defence Science and Technology Laboratory (Dstl) has worked with a QinetiQ-led industry team – also including BAE System, Thales and SeeByte – to develop an open information architecture, known as the Maritime Autonomous Platform Exploitation (MAPLE), to enable the effective and efficient mission planning, integration and control of unmanned vehicles, and exploitation of their capability into the chain of command.

The central aim of MAPLE is to ensure that the operation of multiple maritime autonomous systems in a squad or swarm does not generate the attendant need for multiple operators per vehicle. To achieve this, the MAPLE information architecture has been developed to address the need for increased integration and autonomy such that a single operator, situated within an operations room, can plan, task and manage missions involving multiple vehicles. It is agnostic to different vehicle types – air, surface and underwater – and its modularity means it can accept different or new 'apps'.

The other principal driver for MAPLE came from a recognition that the life cycle of unmanned systems is significantly shorter than that of their host platforms. Consequently, reducing the burden of integration and increasing the commonality and durability of interfaces is highly desirable in order that rapid update cycles become both achievable and affordable.

DEPLOYMENT OPTIONS AND PLATFORMS

The move away from dedicated MCMVs and towards a scalable, reconfigurable and easily portable 'off-board' toolbox opens up new possibilities with regard to the deployment and sustainment of MCM capability. For example, routine route survey missions in estuarine and inshore waters, or harbour protection work, may now be conducted from a shore site, with modularised equipment transported by land or sea as required. Equipment can also be air-freighted in containerised form to overseas locations to meet a regional contingency, or alternatively pre-positioned in theatre.

Small craft, such as work boats and rigid-hulled inflatable boats, can be adapted to deploy UUV/AUV systems. More lightweight vehicles can be deployed and retrieved by human operators; larger systems will require specialist launch and recovery equipment.

More complex and enduring MCM missions will require some form of larger, self-sustaining seagoing platform from which to operate the toolbox independently at range. One option is to integrate containerised mission equipment and support facilities on commercial vessels of opportunity, such as offshore supply vessels, or on large naval platforms like amphibious vessels or expeditionary support ships. The alternative is to procure dedicated MCM 'mother ships'. As well as embarking different elements of the toolbox, these bring their own organic situational awareness, self-protection, and command, control, and communications infrastructure, together with more bespoke facilities for the launch, recovery, support, and maintenance of different off-board systems.

As early adopters of the MCM toolbox model, the Belgian Navy and the Royal Netherlands Navy are jointly procuring twelve new MCM vessels – six for each navy – which will be used to host a variety of off-board systems. Belgium Naval & Robotics – a consortium formed by Naval Group and ECA – was contracted for the supply of a complete solution comprising both MCM toolboxes and mother ships in May 2019.

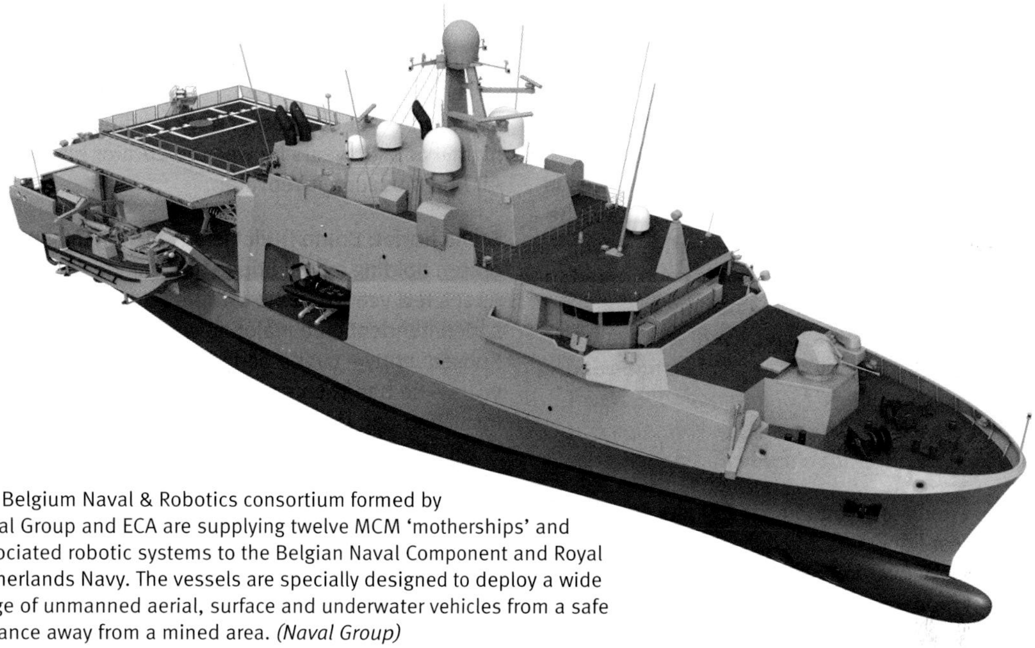

The Belgium Naval & Robotics consortium formed by Naval Group and ECA are supplying twelve MCM 'motherships' and associated robotic systems to the Belgian Naval Component and Royal Netherlands Navy. The vessels are specially designed to deploy a wide range of unmanned aerial, surface and underwater vehicles from a safe distance away from a mined area. *(Naval Group)*

The c. 80m, 2,700-ton displacement steel hull design, to be built by Kership in north-west France, provides an illustration of a contemporary, purpose-designed mother ship integrating combat systems, mission management, communications, and unmanned/autonomous systems.[1] Although designed for stand-off operations, the mother ship will incorporate a level of signature reduction and shock resistance (reflecting the reality that the boundary of the mine danger area will always be uncertain). One notable feature of this mother ship design is the incorporation of a new USV launch and recovery system developed by Naval Group. The two systems – one to port and one to starboard – each consists of a pivoting A-Frame davit, with the USV housed within a floating cradle. Tank testing undertaken at MARIN test facilities has demonstrated the launch and recovery of USVs in simulated conditions of up to Sea State 5/6.

As an interim or 'bridge' step, some navies are looking to repurpose existing MCMVs to operate with autonomous systems, specifically UUVs/AUVs roled for stand-off mine detection and classification. However, the small size of these legacy hulls necessarily limits the number of toolbox components that may be embarked and supported.[2]

CONCLUSION

There is a broad acknowledgement that MCM is on the cusp of a transformation. While current MCMVs are highly capable, they are also slow, expensive and must put themselves and their crews at risk by sailing into mined areas to prosecute their mission. The transition to the toolbox model based on maritime autonomous systems promises to deliver MCM capability in a more efficient, agile, sustainable and safer fashion over the long term.

Notes

1. Kership is a joint venture between Naval Group and Brittany-based shipbuilder Piriou.

2. Some navies appear to believe there is still a longer term need for MCMVs that will continue to have the option to operate safely 'within the minefield'. These vessels will be able to deploy the new 'toolbox' of MCM equipment whilst retaining the low signatures and shock resistance of traditional MCMVs. For example, the Italian Navy's planned *Cacciamine di Nuova Generazione* (CNG) programme for a new generation of coastal and oceanic MCMVs appear to be based on this premise.

Contributors

Richard Beedall: Born in England, Richard is an IT consultant with a long interest in the Royal Navy and naval affairs. He served for 14 years in the Royal Naval Reserve as a rating and officer, working with the US Navy and local naval forces in the Middle East and around the world. In 1999 he founded *Navy Matters*, one of the earliest naval websites on the Royal Navy. He has contributed to *Seaforth World Naval Review* since the initial 2010 edition, and has written extensively on naval developments for many other organisations and publications, including *AMI International*, *Naval Forces*, *Defence Management* and *Warships IFR*. He lives in Ireland with his wife and daughters.

Norman Friedman: Norman Friedman is one of the world's best-known naval analysts and historians and author of over forty books. He has written on issues of modern military interest, including an award-winning account of the Cold War, whilst in the field of warship development his greatest sustained achievement is an eight-volume series on the design of different US warship types. Amongst recent works, *Winning a Future War: War Gaming and Victory in the Pacific War* stands out as a notable contribution to naval history. His latest book, *British Submarines in the Cold War Era*, is the second part of a two-volume history of Royal Navy submarines that forms part of a wider series on British warship types. The holder of a PhD in theoretical physics from Columbia, Dr Friedman is a regular guest commentator on television and lectures widely on professional defence issues. He resides with his wife in New York.

David Hobbs: David Hobbs is an author and naval historian with an international reputation. He has written twenty three books, the latest of which *Taranto and Naval Air Warfare in the Mediterranean 1940-45* was published by Seaforth in 2020. He has also written for several journals and magazines and in 2005 won the award for the Aerospace Journalist of the Year, Best Defence Submission, in Paris. He also won the essay prize awarded by the Navy League of Australia in 2008. He has lectured on naval subjects worldwide and has been on radio and TV in several countries. He served in the Royal Navy for 33 years and retired with the rank of Commander. He is qualified as both a fixed and rotary-wing pilot and his log book contains 2,300 hours with over 800 carrier deck landings; 150 at night. For eight years he was the Curator of the Fleet Air Arm Museum at Yeovilton.

Bruno Huriet: Bruno Huriet is a French merchant navy officer, holding a dual bridge and engine certificate. After a few years on bulk carriers he worked as ship superintendent on a major naval export contract. Subsequently – over a period of more than twenty years – he has been employed by companies in insulation and interior outfitting for both civilian and naval projects. Having worked on naval projects in different countries, he is particularly interested in comparing the varying design solutions adopted by different navies. A long-standing pictorial contributor to *Seaforth World Naval Review*, this is his second chapter for the series.

Mrityunjoy Mazumdar: Mr Mazumdar, who studied applied physics and mechanical engineering, has been a regular contributor to *Seaforth World Naval Review* since its inception. His interests are the sea services of South and Southeast Asian countries, as well as the lesser-known naval and air forces around the world. His words, pictures and research have appeared in many naval and military aircraft publications including *Jane's Navy International*, IQPC's *Defence Industry Bulletin*, *Shephard Media*, *Ships of the World*, *Warship Technology*, *Air Forces Monthly* as well as the standard naval reference books. Having grown up in India and Nigeria, Mr Mazumdar lives in the 'wine country' north of San Francisco with his wife.

Nikolai Novichkov: Nikolai Novichkov is a well-known Russian analyst specialising in aviation, naval and missile technology. Dr Novichkov heads the Information and Analytical Department of the United Military Edition of the TASS news agency. He is the author of many articles, books, reference works and dictionaries and has regularly collaborated with 'Western' publications such as *Jane's*, *Aviation Week and Space Technology*, *Naval Forces* and *European Defence Review*, amongst others. In recent years, he has written a detailed, two-volume analytical book, *Russian weapons in Syria: analysis, results, conclusions*, as well as *Combat use of unmanned aerial vehicles in Nagorno-Karabakh*, both published in Moscow.

Richard Scott: Richard Scott is a UK-based analyst and commentator who has specialised in coverage of naval operations and technology for over 25 years, with particular interests in the fields of naval aviation, guided weapons and electronic warfare. He has held a number of editorial positions with Jane's, including the editorship of *Jane's Navy International* magazine, and is currently group Consultant Editor – Naval. Mr Scott is also a regular contributor to other periodicals, including the *AOC Journal of Electromagnetic Dominance* and *Warship World*.

Alejandro Vilches: Alejandro A Vilches Alarcón is a naval architect from Cádiz University. He is currently employed as a ship manager at Gibdock, a shipyard located in Gibraltar. He has previously worked in a variety of shipyards in Spain, France and the United Kingdom on both civilian and military projects. He began his writing career in 2002 and has written articles for specialised magazines in Spain, France, the United Kingdom and the United States. Alejandro has also had four books published in Spanish about the history and development of Soviet submarines and Second World War naval affairs.

Conrad Waters: A lawyer by training but banker by profession, Conrad Waters was educated at Liverpool University prior to being called to the bar at Gray's Inn in 1989. His interest in maritime affairs was stimulated by a long family history of officers in merchant navy service and he has written on naval affairs for over 35 years. This included six years producing the 'World Navies in Review' chapter of the annual *Warship* before assuming responsibility for *Seaforth World Naval Review* as founding editor. In 2020, he was appointed Joint Editor in Chief of the Mittler Group journal *Maritime Security & Defence*. He has also edited *Navies in the 21st Century* – shortlisted for the 2017 Mountbatten Maritime Award – whilst his latest book – *British 'Town' Class Cruisers* – was published in 2019. Conrad is married to Susan and has three children: Emma, Alexander and Imogen. He lives in Haslemere, Surrey.